Metal Detecting

A Treasure Hunting Guide Book for Beginners

(The Ultimate Guide to Metal Detecting and Finding Treasure)

Michael Fujita

Published By **Jordan Levy**

Michael Fujita

Metal Detecting: A Treasure Hunting Guide Book for Beginners (The Ultimate Guide to Metal Detecting and Finding Treasure)

ISBN 978-0-9958659-8-3

Legal & Disclaimer

The information contained in this book is not designed to replace or take the place of any form of medicine or professional medical advice. The information in this book has been provided for educational and entertainment purposes only.

The information contained in this book has been compiled from sources deemed reliable, and it is accurate to the best of the Author's knowledge; however, the Author cannot guarantee its accuracy and validity and cannot be held liable for any errors or omissions. Changes are periodically made to this book. You must consult your doctor or get professional medical advice before using any of the suggested remedies, techniques, or information in this book.

Upon using the information contained in this book, you agree to hold harmless the Author from and against any damages, costs, and expenses, including any legal fees potentially resulting from the application of any of the information provided by this guide. This disclaimer applies to any damages or injury caused by the use and application, whether directly or indirectly, of any advice or information presented, whether for breach of contract, tort, negligence, personal injury, criminal intent, or under any other cause of action.

You agree to accept all risks of using the information presented inside this book. You need to consult a professional medical practitioner in order to ensure you are both able and healthy enough to participate in this program.

TABLE OF CONTENTS

Introduction

The gold rush in the nineteenth century marked the moment people were fascinated and fascinated by finding treasure and also the dream of finding treasure in the underground. This type of activity requires little of an investment, and it can be extremely rewarding for those with a bit of time. If it is successful, the results are only an example of many benefits that you will discover when you begin metal-detecting as a pastime.

The idea of finding hidden treasures beneath our feet is an extremely stimulating thought that has inspired numerous people to engage in metal detecting and searching for treasure buried as a legitimate pastime. The concept has spread enough that now treasure hunting is seen as an actual pastime and people who participate in it are known as metal detectorists. Metal detectors should not be considered a pastime for people who are unemployed. It is used frequently in investigations of forensic nature conducted by the military, police and even in fields like archeology.

I hope that I've attracted your attention to the field. Therefore, with no further delay we can get going and learn more about the fascinating metal detection world!
We're off!

Chapter 1: Is An Introduction To Metal Detecting

Metal detecting is defined as the process of finding valuable metals by making use of a primary instrument called a metal detector. Metal detecting can be a fun pastime, even though it is challenging at times. The most valuable objects found in metal detecting consist of Gold as well as Quartz, Magnetite and Diamond, as well as Silver. Metal detectors were initially created to detect Gold however today they can also detect other valuable metals.

This book I'll provide the information you require to get ground detection, the most effective locations for metal detecting, the different types of detectors to detect various kinds of objects, ground assessment prior to detecting, and more. I will also provide you with information with the equipment you'll need to start metal detecting with success. There are a lot of things to consider. I'll cover them part by part.

Metal Detecting Theory Basic Should Know

The term "theory" is a reference to the principles upon which the execution in an exercise is founded. It is important to understand this concept to aid you with your treasure hunt. If you don't have the fundamental concepts for a specific area in which you are interested, you might get it right in the conclusion.

Without a metal detector, there is no metal detector. Metal detectors are the primary instrument. Metal detectors are an electronic device that is designed to locate metals that have been that were buried long ago beneath the ground , and could include artifacts/relics, silver, gold and coins from the past that are valuable. When a detector detects any of the precious metals, it will notify you that can be in the form of vibrating or audible beeps.

Metals with value are often found on beaches and oceans, rivers and woods. Parks, forests rock, abandoned houses and places where wars took place long ago. My experience has shown me that these locations through the years are excellent places to search for metal treasures.

A metal detector consists by a handle in metal and the sensor. These two components comprise the two main components of the device. The sensor performs the main job. If you sweep across the ground in search of metals, if the sensor detects any metal can be sensitive, it emits a sound. This signals that there's a metallic object that is hidden beneath the ground. For more sophisticated metal detectors, each metals produce different sound.

The distance between the metal object and the detector is an effect on the frequency of beeps. The closer to the target is to the sensor, the louder the beeps. The further between the metal target and the detector, the smaller the sound of beeps emanating from the sensor. Due to this fact, you need required to pay focus on the sound of beeps when you look for metals to ensure that you don't lose great discoveries. Wearing a headphone is also beneficial.

The Short Working The principle of Metal Detector

It's not a good idea to use metal detectors without understanding the principles behind its operation. So, let us look into it. Metal detectors work by electromagnetic effects. The electromagnetic field that is generated by the coils of the metal detector is sent to the ground when looking for metal. When the field comes across an object of metal the field of electromagnetic energy from the coils stimulates the target and the object becomes charged and thus transmits its own magnetic field. When this happens the detector's is able to receive the retransmitted electromagnetic field , and you, as the user of the detector detects a sound that indicates that there is a metal.

My journey towards Metal Detecting Hobby
My dad was an avid metal-detecting enthusiast. He would often find some treasures during the time he was engaged in the sport. As a child, I was not happy watching my dad go home when it was time to finish his work. However, due to the fact that he enjoyed metal detecting and metal detecting, I had no other choice than to let him pursue what he was passionate about. When he was away the time came to go on a the hunt for treasures in the morning , and at night. The man was an accomplished metal-detecting enthusiast because he had several good discoveries.

As I grew in age, I often used his metal detector to look within the surroundings we were going to leave as he showed me how to operate the machine. Since I was a newbie in the world of practice at the time (about 16 at the time) I didn't discover anything of value. But I was determined to not quit.

The actual event happened when I was just 18 years old. Dad took me out into the forest to do metal detectors. At that time I was a teenager, he had bought the metal detector for me. I'm not sure who made the of that detector, however, it was a good one due to its sensitivity to metal substances.

The day that was a sacrificial one, when we arrived in the woods, my guide accompanied me to the area I had to go to, while he was in the same area (not that far away from me). While I was conducting my way to a location and my metal detector went off. The sound echoed in my ears as if it were actual, however, I put my coil of search on the area again, and I was able to hear the audio. I was ecstatic, however, I thought "what could it mean if what's under the ground could be simply a scrap of metal"?

I found the courage to get started and began digging the earth. I acted with the caution that my father had taught me to avoid to cause harm to the object. What do you think I found beneath the soil? I found a relic with historic and cultural value. My dad was very happy with me.

After our trip at home, after we returned He cleaned up the treasures that we had gathered from the day. He was unaware of how valuable the object I found was until he set out to sell it , along with other items. He made some cool cash from the finds and purchased something for me with part of the proceeds. Since that day my enthusiasm for finding treasures grew. I am confident in telling you that I'm a successful metal detector enthusiast.

Metal Detectors are of various types

Metal detectors of any kind you've seen or are planning to purchase falls under one of the three major types of metal detectors. In this article I will guide you through these three types of systems and how they perform best.

The three most common kinds of metal detector systems are:

1. Beat-frequency Oscillator metal detector

2. Metal detectors that are very low frequency and

3. Pulse Induction Metal detector

What is the beauty of Metal Detectors that Beat-frequency-operate?

What I mean by this subheading is what I consider to be the good aspect of BFO Metal detectors. The advantages of metal detectors manufactured using this technique are as follows:

1. Metal detectors with Beat-frequency Oscillator are not expensive

2. Metal detectors that are manufactured using BFO technology are easy to operate

Metal detector with Very Low Frequency

Another type of metal detector, based on the technology that is used to create it. Metal detectors that operate at very low frequencies are the most sought-after metal detectors that are available on the marketplace in the present. This detector is where you'll be able to find treasures made of metal. Metal detectors with this feature give more precise results over BFO metal detectors that have this feature as a disadvantage. Metal detectors that operate at very low frequencies are also known as their abbreviation VLF metal detectors.

VLF metal detectors called induction balance. The detector is constructed with two main coils. The coils used include receiver and transmitter coils. The transmitter coil is the outer coil of the detector. The coil is composed of electrical wires. In the coil of electric wire, electricity is delivered in one direction at first, and later in a different direction. The wire coil hundreds of times per second.

In contrast it is also the internal coil inside the detector, which is constructed from coils of wire. The wire coil functions as an antenna that picks up frequencies and amplify them which originate from the metal treasures that are buried in the ground.

At lower frequencies detectors made of metal are able to detect valuable targets made of metal. This is due to the fact that low frequencies penetrate into the ground more deeply than high frequencies.

Metal detectors equipped with very low frequency technology have characteristics of discrimination when searching for treasures. This means that these kinds of detectors for metal can differentiate between the desirable metals and those that are not required. Metal detectors with very low frequencies are able to identify silver, gold and copper that are in the soil. When it comes to detecting the length of the object, VLF metal detectors can locate treasures as deep as 15 centimeters deep in the earth.

Pulse Induction Metal Detector (PI)

This kind of detector is based on the pulse effect. In detectors for metal that use this method the powerful, brief flashes (pulses) in current is sent out through the wire. Each pulse creates short magnetic fields. Metal detector PI has a one coil that functions as the transmitter and receiver. These detectors are great to hunt for treasures in mineralized ground. If you are planning to go hunting in these locations choose this type that has a metal detector. This kind of detector comes with a sound discrimination feature. The metal detector is a good choice for hunting in saltwater exploration because the metal in this area is extremely conductive in the soil.

Why do people have an interest in Metal Detection?

There are many reasons metal detectorists are involved. There are those who are looking for valuable metals due to their passion for the hobby. Others continue to pursue it for the financial gain. The reason you choose to do it may differ.

A few days ago, I was involved in conversation with Richard an acquaintance of mine and a fellow enthusiast about the subject. I asked him about the reason he's interested in metal detectors. Richard said he enjoys it because every time he comes across tiny golds, his heart is filled with happiness. He added that it's exhausting, but recollecting the things that lie ahead makes him feel good.

Richards goal has always been gold. Because of this Richard is always looking for metal detectors that focus on gold as the primary metal of his interest. This means that there are many different detectors that target different metals. If your goal is silver or diamonds or silver, you may ask the seller of the detector prior to purchasing, but I'll provide some information in the next paragraphs.

Metal detecting is a popular hobby in order to earn a few dollars from their discoveries. This is among the primary reasons that detectorists in the United States and United Kingdom are interested in metal detection. The money that comes by metal treasures may be a source of motivation dependent on the way in which you value the precious metal. Coins that are old and valuable will bring smiles to your face.

To make the most of your metal-detecting outing, consider these suggestions:

1. Find out about metal detection using the right equipment
2. Practice good listening culture
3. Find your target, and, based on that, choose the most reliable metal detector.
4. Know the area you intend to explore
5. Study map of location
6. Learn about the past of the location you are interested in
7. Do not cross the border.
8. Pay attention to safety measures
9. Find long-term search terms

10. Place your detector as close as you can to the surface of the ground

11. Hunt after rain when ground is still wet

Chapter 2: Understanding The Metal Detecting Lingo And Guide

Every profession has its own language they use to talk to one another. The case of metal detection isn't going to be any different. Due to the fact that metal detecting enthusiasts are in the field, these terms are still being used by treasure hunters.

The term "lingo" simply refers to terms metal detector enthusiasts employ among themselves. Sometimes, they are aware of this within themselves. When they talk in certain situations, a stranger hardly comprehends their messages. If you are a beginner is to be knowledgeable the terms. You could be in the company of experienced metal hunters later on. You're sure it'll not be pleasant to be lost among them.

Due to their importance due to their importance, I will walk you through the basics of the most important terms.

14

However, if you don't know I've mentioned a some of these terms in the first chapter. Some of them are BFO that stands for Beat-frequency oscillator metal detector, VLF, which stands for Very Low Frequency metal detector and PI, which is a reference to pulse Induction Metal Detector.

Let me define terms in metal detecting to you:

All metal

This is a term that applies to any the metal detector. If you set the metal detector on to the all-metal mode indicates that the discrimination function of the detector won't be able to function. A metal detector that is in all metal mode will detect every metal object it encounters during the hunt, even objects of junk.

Tone ID

The tone ID can also be an lingo used for metal detectors. The tone ID is an audio tone that every metal that is detected is identified by. For instance, Gold may have high tone ID while Iron artifacts have the low-tone ID. Tone ID may differ based on the specific material. However, with patience and diligent study, you'll be able to know the tone ID for silver, gold, diamond and many other metals.

Visual ID

Visual ID is a feature of a metal detectors that permit you to visualize the metal that is detected in the soil before you begin digging. This feature is incorporated into modern-day made metal detectors because of the advancements in technology. It's a great benefit for us treasure hunters as it has enabled us to acquire higher-value metals, instead of spending our time digging for garbage.

Choppy

This is the sound that an audio metal detector's signal when it detects a metallic that is nearly filtered out by setting mode on the device. The sound is not entirely reliable However, it is suggested that to continue digging to discover the actual substance.

Find

It's the treasure you find in metal detecting that's worthy of keeping. It could be a gold necklace rings, necklaces, or other metal treasure you wish to keep. I frequently use the term "gold" when I am on a treasure hunt.

Low tones

This is the low-sound metal detectors make when they discover precious metals with a low conductivity properties. Examples of such metals include the gold, and also pull tabs.

Bling

This is the term we use to describe extravagant jewelry. The metal could or might have precious elements. If you spot this metal initially, you'll likely be confused as to its nature due to its appearance. The most beautiful bling is jewellery that is filled with precious metal.

Black dirt

This type of sand is that is typically found in old sites. If you dig a hole when metal detecting and are within a few feet of sand, you're likely to uncover at minimum an undiscovered treasure. If I am at the lookout for treasure, when I discover this sand in the course of my exploration I'm generally happy. Black dirt is the most fertile soil.

Sand that is black

Black sand is another kind of sand designated for something beautiful. When you see black sand in this valuable activity, it's an indication that you're near to the gold. Sand that is this type of is high with iron particles.

Bucketlister

Do you know how the word is pronounced? If you look at the word carefully you'll discover that it is an amalgamation of the two key words "bucket and "lister". A bucketlister is an amazing discovery you made while metal detecting. It is truly amazing. You may have never ever thought about finding such a discovery in your lifetime, but it was discovered in the span of a single day.

Cache

A cache is basically an assortment of jewelry or coins that were buried by a number of people at a certain location long ago. Metal detecting is a person who finds these coins earns real profits from these treasures. Sometimes, they are stored in containers. They are also near in relation to the place they are buried. If someone tells you that they discovered a caches, do you understand what cache is?

Cache hunting

The term above was created from two key words "cache and hunt ". In the context of that, cache hunting is a treasure search for coins or other past treasures which are scattered. When it comes to this type of hunt an approach that is specific to the hunter and a thorough study is used to achieve the result you want.

Coinball

Coinball could mean something different to different people based on the area in which it is employed in. In the field of metal detection, coinball can is a term that means something different. Coin balls are dirt that contains coins. If you're looking for something and your machine is ringing simply hold it in place and discover what's within the dirt. There could be valuable coins within.

Canslaw

It is defined as a collection of Aluminum cans scattered over an area due to the fact that they were struck by a lawnmowers. They make detecting metal within the region difficult since the Aluminum that are of various dimensions give a variety of signals.

Color

In the hobby of metal detecting We employ color to define gold. If you're searching for an area in particular and claim to have discovered colors that are treasures then you've found Gold.

Clad

This is a sought-after piece of currency in the absence precious coins of the past. These are brand new coins made using non-precious metals. They are made to be Silver colour in the United States.

Coin Spill

It is a petty problem detectorists often face even although it isn't the case often. Coin spill occurs the event when a coin falls from the bag of a hobbyist following or while detecting. It is usually due to the negligence that the person who is a treasure seeker. To avoid this embarrassing experience, always use a an adequate and well-stocked bag. It's safer to be cautious rather than lose the value that you have discovered following a long day.

Pennyweight

Pennyweight is a measurement equal to 24 grams. It is abbreviated dwt. which stands for denarius, an ancient Roman coin.

Pinpointer

A pinpointer is a hand-held metal detector that is very small in dimensions. It is installed inside an unplugged hole to find the targets. If you dig a hole and reach the point that you wish to go to the exact direction of the treasure that you are seeking You will need a the pinpointer. The practice of using a the pinpointer to find treasure is known as pinpointing.

Assay

Also, a term that is employed for metal detection assay is a process in metallurgy which can be utilized to determine the authenticity of treasure. Detectorists utilize it to determine if the treasure they find, be it Gold, Silver or any other item is pure or not. The proportions of precious metal are determined through essays.

Plug

The plug is a hole which is properly and cautiously created during metal detector. A great plug is usually the work of a skilled metal detectorist. The person who finds plugs saves the earth from destruction , and this practice is recommended.

Tot-lot

The name might sound odd to you However, we are interested in finding out what it's about. Tot lot can be described as a play area designed for children as young as. It is designed to safeguard children from injuries and is typically located near parks. It is also a spot to store lost jewelry. Due to this, many metal detector enthusiasts visit these locations.

Relic Hunters

The name suggests they are metal lovers who specialize in finding treasures. For this type of treasure hunter they search in woods or fields and generally concentrate on areas that represent earlier conflicts, like those which witnessed the Civil War in the United States.

Utilizing Your Metal Detector Correctly When you go on a Treasure Hunt

There are a few things to consider when metal detector. Additionally, there are things to consider when you set out to find treasure. Don't think that just the fact that you have a metal detectors that you can make use of it for anything you'd like to accomplish. Don't be afraid to make use of metal detectors. ....

Things you should be aware of:

Moving and searching

How you carry and move your metal detector is vital in your search for treasure. If you are searching for treasures, make sure that you maintain the height of your search coil to 2 inches, and your detector should be in line with the ground to get outcomes. Don't elevate your search rod enough that it is able to establish a connection with the target beneath the ground.

Your Movement

Don't be in hurry when you go out in the field looking for treasures. Make sure to take your time and take your time. If you hurry to get there, you could be sucked away from the treasure that you could have discovered. While you are walking you should move your search device with a speed of around 2-5 inches per second.

Cover Holes

Many metal detectorists are at fault for this. For those who are just beginning there is no need to emulate a bad model. You can just go with the best. If you decide to find treasures using your metal detector and in this process you make holes, make sure you make sure to cover them. This will help keep the land tidy and neat.

Don't Destroy Properties

Being granted the chance to hunt on someone's property is not a right to demolish other valuable items on the land. Don't destroy the plants or animals' lives. Don't destroy any other items that are valuable. If you damage something by mistake, you must let the owner be aware.

Management of trash

In the process of digging up the ground in which your detector has spotted a potential treasure, it is possible of coming across garbage. It is part of the process of detectors and as a result, it is possible to be dirty. When you remove garbage during your exercise, don't let them pollute the surroundings. You can put them in a bag and put them in a nearby garbage bin. Make sure that we keep our country tidy. Be an excellent citizen.

Important Notice Don't go on a treasure hunts on any property you are not allowed to conduct a treasure hunt on. Don't think that every piece of land is free to visit. There is no land that is free. Therefore, you should ask questions prior to searching to find treasure in any area. Be sure to follow the laws that governs any land you come across in any part of the course of your metal-detecting hobby. Avoid paying a costly costs for violating the law.

There's another important message you should listen to. Before you begin metal detecting in any area that is mainly in a bushy area, inform your family members be aware of the directions. There could be dangers for animals in these kinds of areas. Because of that be aware. Also, make sure you ensure that you are properly prepared for any move so that in the event that an issue such as that occurs it is possible to combat the issue you are able to take on. I have killed a dangerous snake on a treasure hunt since I had the equipment to complete the task at the time. Be prepared always.

Chapter 3: The Rules And Permissions For Metal Detecting

A society that isn't controlled through law enforcement and law will not remain. Similar to that, anything that's not controlled by rules could result in our society becoming chaotic. Because of this metal detecting follows rules to guide it. This is the reason why it is known as the most organized in certain regions. The next chapter I'll explain the basics of metal detection.

In addition, getting permission before hunting on any property is extremely vital. It is required in the United States, United Kingdom and Australia. If you don't get permission prior to hunting on certain areas, you could be punished. It's the act of trespassing. This is not your property and, therefore, you need to inquire for permission before you can get it.

Are detectorists punished for not adhering to the rules of metal detection?

Yes. No matter how skilled you consider yourself to be in metal detecting, if violate the rules of metal detectors, you'll be penalized. The fine you are liable to pay could be greater than the value of the findings you find.

It is essential to know the Rules of Metal Detecting Explained
The rules for metal detection are as follows:
Rule Number 1: Do Not Intrude
If you're out to search for metal, be sure not enter. If you find that a certain property is restricted and you are not required to enter the property because you wish to search for precious metals. If it's written that no one is allowed to enter the area to hunt, then please follow the instructions of the landowner.
Rule #2: Do Not hunt in Sensitive Zones
Life is much more valuable than any treasure. Government properties are not to be destroyed at any time, no matter what you do with your discoveries. When searching for treasure, don't hunt through areas that have pipelines buried. If it's pipelines for gas that traverse the ground striking them could put you in danger. Skin can be burned or even die from the impact. Petroleum is extremely flammable and should be avoided. avoid these places.

Beware of places where electrical wires run through underground. The risk of electrocution is high in this area.

Rule #3 Keep clear of National and State Monuments

Don't make the error and take your metal detector in close proximity to any monuments that are national or state-owned including parks. They are protected. Within the United States, a national monument is a protected zone which is akin to an national park, but it can be made out of any land controlled or owned through the Federal government via proclamation by the President of the United States. Don't let the government manage your case, as you'll sweat.

Rule #4: Use prudent caution when digging towards targets

While digging ensure that you use the proper precautions. This is due to the fact that deep-seeking machines are able to detect hidden wires, pipes and other potentially hazardous substances. If you take care when your digging you will be able to immediately alert the appropriate authority when you discover the presence of any.

Rule 5: Avoid Military Zones

This last rule should be a part of your brain. Don't make a mistake and avoid hunting in military areas. Do you know whether dangerous weapons are hidden beneath the earth? Do you wish for your body to be damaged by radiation from dangerous sources? I'm sure you do not like to suffer from that. However, regardless of what you've heard about the potential treasures to be found in these locations Do not visit them.

How do I obtain Permission to detect on A Land?

The right to hunt on land which aren't owned by the hunter is essential. It will provide you with peace of mind and allow you feel confident that you travel towards your treasures. In this subheading , I'll explain how to get permission to hunt at any place you like that isn't an area that is protected.

Find the owner of the land.

It is costly to purchase property. Also purchasing land is costly. Some people need a to save for a long time before they can own their own land. In light of that it is essential to conduct an inquiries about the owner of the land prior to embarking on a hunt. If you have a relationship with the owner personally or have a meeting with him, you may meet the person and have a discussion. If he tells you that you'll provide him with a little of the items you finds after you have hunted and you be in agreement with him as long as you're happy with the decision. You could start by talking to the people you already know prior to the ones you don't have a relationship with.

Don't go with Your Machine

There is no need to tell the public that you're going to a residence in order to ask that they allow you to hunt on their land. Are you aware of the possibility that your neighbor doesn't like your manner of dressing and may ruin the thoughts of the owner after you depart? Be aware when you set out to request permission to hunt within any given area.

However the person you're meeting may not be fully knowledgeable of metal detectors. He may not allow you into the house as he may not know if the item you're carrying could be a weapon that is dangerous. If he does let access, he could not like your presence and could not approve of your request. Keep your detector in the house before asking permission to go metal-detecting in a specific location.

Be Persuasive

Discuss your concerns between the property owner and the person who owns the property land interactive , and be persuasive when discussing with them. Make yourself convincing and act as an expert in marketing. Help him see the reasons that he should let you in on his territory. It is possible to convince him that things can change. Let him know that you might discover treasures that are be worth thousands of dollars, and that he, your own can still benefit from it.

Stick to the Agreement

If, at the beginning, you came to an arrangement with the property owner be sure that you adhere to the terms of it. If you have agreed with him to complete all excavations after digging make sure you do this with diligence. In the event that you agreed that you would offer him a small amount of results, then do so. By sticking to the agreement, you can result in him giving you another great location to search for precious metals.

Chapter 4: Mastering Your Metal Detector

If you don't know how to operate the metal detector, you'll be unable to find good results. This is why you need to know this section. The best end result of any metal detector outing is treasure that could include Gold, Silver, coins jewelry, artifacts, or any other precious finds. With regard to this regardless of how expensive the metal detector is when it fails to produce any finds that is the same as being a less than sensitive detectors. Because the final product that is sound is crucial for us, I must the ability to fully reflect this section.

Since this course is focused on metal detectors and other crucial elements of the machine I will also discuss additional fundamental parts that I have previously discussed. This will help you truly understand the lessons. Without the metal detectors, there would be no such thing as the metal detecting interest of today's. Therefore, let's keep the ball going.

Search coil

It is also the home of the search engine for every metal detector. If there isn't a search coil within the detector then there is no way the detector will be able to communicate with the intended target. The search coil is the main point of contact between the target and the metal detector.

The search coil can be described as a a set of wires near the top of a metal detector that is used to find the targets. It's these coils of wire that make metal detection possible. The coils that comprise the metal detector may be large or small in dimensions. A lot of metal detector enthusiasts prefer smaller coils over large coils, however, each has its strengths and weak points.

Search coils that are constructed using small coils are more sensitive to larger coils for the ability to detect targets. When the target is hidden in dirt, it is difficult to use this kind of search coil and therefore large coils are the ideal choice in such a scenario. It is recommended to have metal detectors made with large and small coils. This will make it simple to change depending on the location you are in.

The 8 inches coil is an all-purpose coil that is the most sought-after coil for metal detectors. For Gold prospecting the 10 inches by 5 inches coils are ideal to find tiny Gold.

Audio Tone

Once you reach a some point in the field of detecting, you'll begin to observe some aspects about the different tones of metals. Metal detecting can be described as an education program where you are taught the basics of what you will learn as you progress within the field. There will be a point at which where you'll be able to hear the sounds of Gold, Silver, Iron and Steel in the event that your detector is able to can detect these substances beneath the ground. When you are at a advanced level of expertise, then you will not have to dig every inch you find. As a novice you are required to be able to do this at the moment.

In general the event that your search coil is detecting non-ferrous metals, it will give moderate to high-pitched sound. This is the way detectors in general are constructed. In the field of metallurgy, a nonferrous metal is a material comprising alloys that doesn't contain iron in significant amounts.

Examples of nonferrous metallics include Gold, Silver and Copper.

If your metal detector comes upon ferrous metals, it emits a low-frequency audio tone. This is because of it you will develop a an excellent listening technique while you search for treasures in various places. The most common examples of ferrous metals include steel and iron. Certain artifacts are created by using steel. Metal detectors is able to locate them and provide the low tone back.

Indicator of the target

The indicator for the target of metal detectors simply indicates the metal that is identified through the device. If it's gold it will be displayed on the indicator for the target that is on display in the detector.

Target ID Number

This feature is available in the latest metal detectors, you will be able to look up the numbers of targets identified on the display of your device. The range of numbers varies based on the model and manufacturer. For some, the numbers range between -4 and 44 on the Xterra 305, and -9-48 on the 505 X-Terra. Negative numbers are ferrous targets while positive numbers indicate nonferrous targets.

Depth Indicator

The depth indicator on metal detector devices is a very welcome development. Thanks to this feature, a metal detectors inform you of the depth at which a target lies below the ground prior to digging. It can be medium, shallow or extremely deep. It will help you decide the amount of energy you need to devote before getting to the treasure. Depth indicator is visible on the display of certain metal detectors.

Magnetic Mineralization Intensity

This characteristic found in some metal detectors alerts the user of the mineral's intensity beneath the surface. This is a significant factor in helping users choose the best metal detectors to be used in a particular setting. One example of a detector with this characteristic is the Nokta Makro Gold Metal Detector.

Learning and Using Your Metal Detector

The first step when you have purchased the metal detector is to open it and put it together. The metal detector is comprised of several parts, such as the detector, control box shaft and the handle. The various parts of the metal detector are arranged into a single device at its end.

If you open the box that contains your metal detector once you receive it, the very first thing you'll get from the package is a manual. In the guide, you'll be shown how to set your machine up. A lot of metal detectors have batteries that power the device however if it is not present in your machine then you can purchase the necessary batteries to power your machine.

After you have put together your metal detector the next thing you need to consider is studying the operation of your machine with respect to various metal kinds. This refers to the sound your detector is able to provide for each type. For instance, the audio for Gold is not identical to that of Iron.

To find out the unique sound to each type of metal, you can bury various kinds of metal in various places within your surroundings. You can sift your detector through all of those metals in order to find out what sounds are unique to each. As time passes, you'll be able to master your instrument.

Do I have to change My Metal Detector Every Now and Again?

The answer is no. If the metal detector you have is not working and doesn't provide the accuracy you want when searching for treasures, you can replace it. If the detector you're using is working it is not a need for regular replacement. A single metal detector that is appropriate for detecting in certain kinds of surroundings is fine. This will allow you to be able to master and "marry" your device.

Troubleshooting in a Metal Detector

Troubleshooting is the process of identifying the solution to the problem that arises when trying to finish a specific task. If you are out on a the hunt for treasure, it could occur a moment when your machine ceases to function. Then you could be upset. Troubleshooting actions to use are:

1. You should check the battery compartment of your machine to figure whether it is properly positioned when the detector is unable to come on.

2. Clean the battery's metal terminals using a clean dry cloth.

3. If your metal detector fails to start, you must replace the batteries, as this could be the cause of the issue.

4. If your metal detector's screen displays OL at the bottom of the screen, that means Overload, look to determine if your metal detector's search is situated in massive metal. If so, take it off

5. If you are experiencing issues with the level of detection on your detector, make sure to examine the settings of your device and then fix the issue.

6. If your search coil's part is not tight, tighten the bolts with plastic but if that doesn't solve the issue, simply replace the washers.

The Effects of Minerals on Metal Detectors

As a novice I would like you to be aware that every metal detector is not appropriate for all soils. There are certain minerals that stop "ordinary" detection devices from detecting certain metals that are found in the ground. Minerals are a naturally occurring crystal which is not able to be broken down into smaller parts. A few of the naturally occurring minerals such as irons, salt and hematite could influence your metal detector when exploring deep within the earth to find the treasure. If you're looking for treasures in the mineralized ground, one the most effective metal detectors to accomplish the task in the zone is the Ground Balance metal detector.

Chapter 5: Classes For Metal Detectors

In addition to the three primary kinds of metal detectors that are depending on the technology involved there are various kinds that metal detectors can be found. They are classified dependent on the maker and their use in relation to specific metals and the environment. The next chapter I'll explain the different metal detectors and also what they excel at. After reading this chapter, it will help you in making the right decision when it comes to choosing the best metal detector. Let us get started.

Minelab Metal Detectors

The Minelab detectors have proven to be among my most reliable as expert treasure hunter. They offer me the results I require at the time when I need to change to a new mode when hunting for treasure. Metal detectors from Minelab are great detectors for Gold prospecting. They are able to detect Gold easily , when the search coil is positioned through the material.

Another advantage of this type of detector is the ability to dig deep into the entire group. Metal detectors developed by Minelab are able to penetrate the ground using the search coils and provide signals back on the objects that were discovered. You can activate the discrimination mode to distinguish the garbage from the useful as you explore.

With this type metal detector, you are able to change search coils based on the location you intend to search for. If you're looking for a bigger coil, it is possible to switch and the reverse. Minelab makes the machine adjustable. Minelab Metal detectors is made by a company named Minelab.

Fig 5: Equinox 800 Minelab metal detector
There are a variety that minelab detectors are available. Metal detectors can be found in:

1. Minelab X-Terra 705
2. GPZ 77000
3. CTX 3030
4. X-TERRA 705 Gold and Dual pack
5. X-Terra 505
6. X-Terra 303
7. Vanquish 340
8. Vanquish 440
9. Vanquish 540
10. Equinox 800
11. Equinox 600
12. E-Trac
13. Safari and many more

Every single one of Minelab metal detectors are constructed using advanced technology to operate to different types of ground conditions, and to make discoveries.

Equinox 800 Minelab's metal detector offers four main modes that you can choose from based on the type of ground and the desired location. The four modes are Park, Field, Beach and Gold. Minelab metal detectors are durable.

Bounty Hunter metal detectors

Bounty Hunter metal detectors are cheap and easy to use metal detectors that work with VLF (VLF) technology. Metal detectors are the ones that many of the professional treasure hunters in the present day began with. Regardless of the ease that the detector is, it has the essential tools you would expect from the standard metal detector.

Metal detectors like this feature discrimination properties. With the discrimination dial you can distinguish trash from the precious metals while you advance in your metal detector. This will allow you to find value instead of garbage in the form of fragments of lead.

Bounty Hunter Metal detector comes with Sensitivity buttons on the box for control. By using this feature, you can boost the sensitivity of the detector when you search for treasures. If you turn the Sensitivity up, the amount the searcher's coil will travel deep into the ground is larger. This means it will be able to detect the ability to detect a greater depth.

Warranty is an approach to keeping a good relationship between businesses and their customers. It's an important aspect to consider when selling electronic goods to customers. Bounty Hunter metal detectors come with a warranty. If your recently purchased Bounty Hunter detector experiences problems within the warranty period the customer can bring it back from the vendor for it to be repaired or replaced based on the severity of damage.

Additionally, Bounty Hunter metal detectors are suitable for children. The company makes a variety of metal detectors that are designed to be used by teens in society. I have a memory of buying an item for my child a long time in the past. He was impressed and made some excellent discoveries using it. One thing I love about this detector is its lightweight. It's easy to carry around by small children.

Metal detectors in this category can be adjusted. They can be adjusted to whatever length you'd like they will be. Perhaps someone else would like to use it but doesn't like the level of adjustment at this point, he can make an adjustment to suit his needs.

One of the problems with the majority of Bounty Hunter metal detectors is their inability to find treasure within mineralized areas. This is a flaw and is a result of the technology that the metal detectors were built on. Bounty Hunter Metal detectors function using VLF technology.

Bounty Hunter metal detectors are numerous on the market right now. They include the following features:

1. Fast track metal detector
2. Tracker II metal detector
3. Tracker IV metal detector and
4. Bounty Hunter pioneer EX metal detector
You can find more Bounty Hunter Metal Detectors via her site www.detecting.com.

Bounty Hunter's Pioneer Metal Detector
I'd like to give more details about this particular class of Bounty Hunter metal

detector because I love its features. Bounty Hunter's first EX Metal detector is lightweight and effective in gold detecting. It can be adjusted and be used by children and adults to find treasures. It is manufactured in the United States of America. If purchased, the product comes with a five-year warranty.

The control box in it houses of the electronic components. This includes sensitivity, discrimination and other characteristics that are based on electronics. When using this machine make sure to protect the box from water to prevent the possibility of damage. This device is powered by 2 9-volt battery. They must be alkaline batteries in order for the machine to work properly.

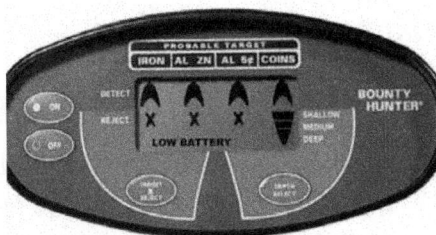

Figure 5.1 Control box for Bounty Hunter Pioneer EX metal detector

By using the Depth Select touchpad located on the control box on the machine you can alter how deep the detector is able in order to detect the ground. Additionally the touchpad that controls depth can be used to minimize electromagnetic interference to the detector made of metal.

The metal detector could find metals composed out of Iron and Aluminum Zinc Gold, and coins. If the target is identified during the search, the machine emits an audible sound. There are three main audio sounds included in Bounty Hunt The Pioneer EX metal detector, however one of them is played at a specific point.

With the touchpad TARGET REJECT you can choose to discard any metal that you don't wish to come upon any. It makes your job easier and prevents you from wasting time with unwanted metals.

Teknetics Metal Detectors

It is a great beginner's type that includes a metal detector. It is easy to use and move around. There are numerous great features that are built into this detector that allows it to be used at various places.

The Teknetics metal detectors were produced by a firm called First Texas Products. The company is located situated in the United States of America. They manufacture Bounty Hunter Metal Detectors, as well as Fisher Labs metal detectors.

Can Teknetics metal detectors find Gold?

Yes. Metal detectors from Teknetics can detect Gold. Regardless of the price tag for the device it still has the ability to find Gold while hunting.

Is Teknetics metal detectors good for beginners?

Yes. This metal detector is ideal for those who are just beginning. In fact, it's among the top detectors for novices due to its simplicity.

Fisher Lab Metal Detectors

Fisher Lab metal detectors are manufactured through First Texas Products Company as described previously. The detectors they make are lightweight and easy to operate and also. They are Fisher Lab metal detectors, they have ones made for multi-purpose metal detectors for treasure detection. Examples of such detectors include Fisher F44, which is a Visual and Music Target-ID Metal Detector Fisher F22 Audio & Visual Target ID Metal Detector Fisher F11 Audio & Visual Target ID Metal Detector Fisher F2 Audio & Visual Target Identification Metal Detector, and Fisher F4 Visual and Audio Target ID Metal Detector. These metal detectors are effective at detecting jewelry, coins and objects of the past.

Fisher Lab metal detectors are constructed with sensitivity to sound and tones that discriminate. The tones produced by these detectors vary based on the metal being detected. Metal detectors are manufactured by the company that identify primarily Gold. The detectors made of metal are Fisher Gold Bug-2 Metal Detector and Fisher Gold Bug Metal Detector. The detectors work with 2 Search modes which include: The entire Metal as well as Discriminate. When you purchase them, they will come with a an instruction manual that will guide you through.

Garrett Metal Detectors

Garrett metal detectors can be described as basic and a great detector for those who are new to the field. They come with sound capabilities. The control box can be used to control the functions of the detector prior to and after the detection. It is possible to control the sensitivity and discrimination settings of the machine from the touchpad. The detectors of this category come with an adjustable strap for the arm and cuff. They come with a an instruction manual that will help you to put it together. it.

There are numerous Garrett detectors on the market. They include Garrett APEX metal detector, Garrett ACE 250, Garrett AT Gold, Garrett AT Pro metal detectors as well as Garrett ACE 350 metal detector, which are just a few of the numerous.

Certain Garrett metal detectors are control boxes that are waterproof. Some examples of such metal detectors include Garret AT Gold and Garrett AT MAX.

To find a wide selection of Garrett detectors, go to the website of the company at https://garrett.com.

Garrett ACE 250 Metal Detectors

Garrett ACE 250 is a well-built machine, and weighs around 2.7bls. It comes with an underwater search coil. The metals detected by this machine are jewelry, coins, Gold, Silver, Relics, and more .Garrett The ACE 250 metal detector can be used with headphones . However, I recommend to purchase a different model that can provide more quantity in the way you prefer it.

The machine has proven itself to be an excellent metal detector for novices. It's sturdy, easy to operate and extremely affordable.

Going hunting in the woods with Garrett ACE 250 Metal Detector as a beginner

If you're looking to hunt using this type of machine, especially as a novice I suggest starting with a sandy area. It is possible to begin with a the beach treasure hunt. This is because searching on this kind of spot is simple. It also helps you learn how the detector operates to give you the best results.

Garrett ACE 250 is a excellent metal detector. You will be happy when you learn the techniques to operate it.

Chapter 6: Security And Relics Detection

The phrase "safety first" is not new. If they're new to you, they have heard it since high school until now. Safety is a must. It's a foolish decision for a metal detector enthusiast to find amazing finds and then not spend the money earned out of the finds due to unsafe measures. Due to the importance of security is when it comes to metal detecting, I'll go over this in the next section. This will allow you understand what to do before and during your treasure hunt.

Relics are valuable treasures. Relics are defined in this way as objects which are from a previous time and, in particular, one of historical value. Museums love relics, and will be willing to pay a substantial amount to possess them. They are cultural treasures.

There are several relics that were created from costly Gold and have not been discovered because of the impact of wars on these locations. Some are of significance to the royal family. Imagine finding a piece of artifact which contain a significant amount of Gold? Even if museums don't make a significant amount businesses and individuals can purchase and then duplicate other objects made of metal that could make them more money in the end.

In this chapter of the book I will also go over the process of finding relics. I will provide you with information the top metal detectors that you can employ to search for Relics.

Safety Tips for Metal Detecting

These are the safety steps you should take while metal detecting

1. Make sure you wear the correct clothes

It is crucial to put on appropriate clothing to safeguard yourself when you go out to metal detectors. The location you're in will determine whether you dress in lightweight or heavy clothing. Also, make sure you have the appropriate footwear and safety footwear. Don't think that you're simply out there to play, so take precautions to protect yourself. Also, make sure you check the weather forecast to determine the weather forecast for that day. It will let you determine the weather conditions for the day, and thus help you dress accordingly.

2. Be aware of sunlight

Many metal detectorists aren't as concerned about this. I'd like to bring your attention that the incidence of skin cancers in the present day society is very high. The cancer can be caused due to Ultraviolet radiations from the sun. You can also examine the weather report for the region you're planning to search for treasure and discover how bright the place is likely to be.

You should also wear sunglasses when you begin to hunt for treasures. If you don't have one, you should consider purchasing an item for your own. Be sure to protect your eyes, as they are very important.

3. Use Knives

You should be prepared as you seek out treasures in certain regions. You don't know the dangers of hunting a wild animal that you're planning to enter and hunt. Therefore, you should use sharp knives as a safety measure.

4. Keep first aid supplies on hand

Have you ever thought about sustaining injuries while searching for treasures? It can happen, and that's why you must be prepared to be ready at all times. I have a close friend who erroneously hit a the shovel with his leg while he was searching for treasure. If something like this occurs to you, and you've got first-aid items in your backpack You can provide yourself treatment that can sustain the victim until you have finished your hunt.

5. Keep in touch with your contacts

When you are on a metal-detecting, make sure to go using your cell phone. Always keep in touch with your family members about your location and where you are about. This will make it easier to reach you at any time.

6. Maps are a great way to hunt for treasures

Regardless of the fact our electronic devices like phone may help us find the place we'd like to explore and find treasures, it's not enough. This kind of electronic device is not always effective in certain situations. Due to this problem, it is recommended to travel with a physical map to help you navigate the area you're looking for treasures. The map will guide you to the destination you want to explore and assist you in finding the return route.

Relic Metal Detectors

If you want to discover relics disappearing in all environments for years You will require a an relic metal detector to accomplish the task. If you don't have an instrument that is specifically designed to search for relics you can make use of a multipurpose metal detectors and switch it to Relic mode using the controls box.

Relic metal detectors are able to find metals that are composed out of Iron, Steel and Brass. Relics with historical significance are typically made of any of these metals. The first step is making a choice about the area where you'd like to find the treasures of the relics, and the second is to select the most suitable equipment for the task. Metal detectors that have low frequencies are the best to find treasures.

Certain metal detectors that detect relics include:

* Bounty Hunter Platinum metal detector
* Fisher F5 metal detector
* Garrett ATX metal detector
* XP DEUS Wireless Metal Detector
* Minelab CTX 3030

* Teknetics T2 Special Edition
* Garrett ACE 400
* Nokta Makro Invenio and
*Whites MX Sport

In the above list that includes metal detectors you are able to select any of them for relic detection. Since Relics are found at a lower frequency, you can lower frequencies of the detector once you discover.

Garrett ACE 400 for Relics Metal Detecting

Garrett ACE 400 is a modern Garrett Metal Detector. It's built with modern features, which is the reason it can detect the relics. The capabilities included in this machine are Iron Audio, Digital Target ID and Frequency Adjustment.

Utilizing this Frequency Adjust option, it could assist you in finding deeper and find more treasure, and less. Additionally, this property allows you to decrease the frequency in order to find Relics. Relics detection is higher in low frequency.

Chapter 7: Coin Shooting And Metal Detectors For Other Areas

"Coin Shooting" is the term used to describe "coin shooting" is the practice of looking for old coins by using an right metal detector. I can remember my first attempt on detecting coins. It was not an easy task because I searched for hours before I made the discovery. Do you know what? When I discovered an old Roman coin in a spot that I was extremely excited. The excitement led me to become addicted to the shooting of coins. I am a fan of this pastime. An old-fashioned coin you can find may be amount to up to a thousand dollars.

It is essential to know your metal detector to perform in the field of coin shooting. The metal detector must be set up in a manner that it is able to distinguish Iron. The most common target is Silver. When you discover old Silver coins dating back over 150 years old You will be smiling. You'll be smiling because you are able to the opportunity to get your hands on a precious treasure.

Metal detectors are best suited for coin shooting. They were developed by the manufacturers in order to provide coin shooters with the highest quality results. Additionally, there are metal detectors with multiple functions that fulfill the same task. With these detectors all you need do is give it the required settings to recognize coins via the control box, and you're good to go.

Metal Detectors for Coins Shooting

Metal detectors that are suitable for shooting coins are as are:

1. Garrett ACE metal detectors
2. Bounty Hunter Gold
3. Whites CoinMaster
4. Fisher F4
5. Garrett AT Pro
6. Fisher F70
7. Whites MX Sport and
8. Nokta Makro Anfibio Multi

Tips on Coin Shooting and Maintenance of Coins

As a veteran metal detector enthusiast I'm required to teach you something I have learned about this area of the most popular hobby around the globe. There are some tips that you must have prior to diving into the world of coin shooting. These are the following suggestions:

1. If you take coins from a location you have found, don't rush to leave because there could be a cluster of coins in the same area. Therefore, be patient and look for any other coins.

2. Dig slowly and cautiously. If your metal detector detects coins, you should slowly and cautiously dig around the area to ensure you don't damage the coins. Take your time when digging.

3. Don't take your time cleaning your coins. If you scrub it too often with a powerful or sharp objects you could damage the surface of your coins. It is best to clean it with water or keep it in a semi-crude form until you can get it to the purchaser.

4. Utilize a pinpointer that is also known as a probe to direct you to exactly that the coins are located in the hole you are digging. Coins are tiny and consequently, you will require this device to set your eyes focus on a specific location.

5. To get the best results for a successful result, you must ensure that your detector can distinguish other metals from coins

prior to shooting. Set your detector in the coin mode.

Which is my nearest location for shoot coins?

There are a variety of places where you can visit to find the coins you want, such as:

1. Homes that were abandoned years ago
2. Regions that were a part of wars many years ago , such as World War 1 and 2 arears
3. Parks
4. There are rivers (only go if you've got prior experience swimming)
5. Beaches
6. Back yards and
7. Ball fields

Underwater Metal Detecting

The most common feature of metal detectors underwater is the search coils that are waterproof. They're made to ensure they won't become damaged.

If the machine is sweeping over the metal while you are in water, it will beep. With the audible sound it is possible to use your hand to scoop the white sand surrounding it and extract the valuable from it. Sometimes, you will need make use of a hand pinpointer to determine the exact location of the desired target. The pinpointers that are used to detect underwater metal have also been made waterproof. Make sure that you are aware of this before you begin your metal detecting underwater you should be able to swim and dress in a proper manner.

Do not bring your standard metal detector that is used for entry into lakes, rivers, or oceans to search for treasures since it is not going to deliver the results you expect. Metal detectors that are underwater are waterproof and are able to be able to withstand stress. Take these tips into consideration.

Metal Underwater Detectors
There are a variety of excellent underground metal detectors. They include the following:
1. Minelab Excalibur II
2. Bounty Hunter Tk4 Tracker underwater metal detector
3. Garrett at prounder water metal detector and

4. Minelab equinox 600 multi IQ underwater metal detector

Metal detectors that are underwater are submersible and completely submerged machines. If you are working with them, they can provide the highest output with the least effort. Metal detectors with multiple frequencies. They can be submerged in the surface of the ocean. They can be utilized in lakes, rivers or the oceans. Other metal detectors underwater that are highly effective are:

* Kkmoon underwater metal detector

The machine is based on Pulse Induction Technology. It is able to detect treasures up to 30 meters deep in the water. It is ideal to hunt for treasure in salt as well as fresh waters.

* RM RICOMAX under water and

* Kuman automatic underground metal detector

All of these machines come with instructions on how to put them together and operate them.

Minelab Excalibur II Metal Detector Underwater

Minelab is well-known for being one of the top metal detectors producers. They produce products of excellent quality and they have demonstrated this through the Excalibur II metal detector. The Excalibur II is thought to be to be the most effective underwater device made by the company at the date of publication in this publication. It's a clear illustration of the advancements in technology on the metal detectors.

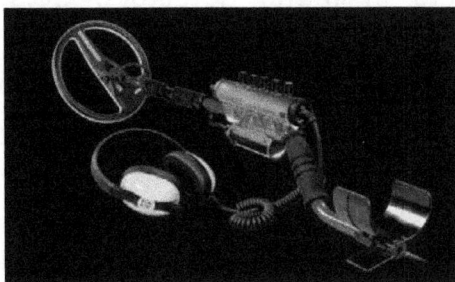

Fig 7: Excalibur II metal detector
The machine has a remarkable underwater performance. It operates using the multi-frequency BBS technology. It's waterproof for up to 200 feet (66m). It is equipped with a thinline 10-" coil. It's not too heavy and is more balanced.

As far as technology is concerned, this machine operates using Broad Band Spectrum (BBS) that ranges from 1.5 kHz to 25.5 kHz. Excalibur II metal detector functions with a rechargeable battery that will last for a uninterrupted 12 hours when fully recharged. If you're an underwater diver who can detect the water, you could consider it.

Gold Detecting

Gold is typically the primary focus of all beginners in metal detectors. Gold is a valuable metal, and everyone would like to get his hands on gold. It is available in chunks or lumps. Certain jewelries are made from gold, which if they are lost and later found by a metal detector enthusiast, could be a source of income for the person who finds it. In the jargon of metal detecting, Gold is referred to as color. Pure Gold is slightly reddish yellow in hue, however, colored Gold with different shades can be made.

Where is it possible to find Gold be found?

Metal detecting is a process where there are locations that Gold can be found. These are the locations:

1. River
2. In waterways
3. Ocean
4. Rocky regions and
5. In the gold mining areas

Metal Detectors for Gold Detecting
Although almost every metal detector can detect Gold but there are some which are more effective. They are made of metal to serve the primary purpose of Gold detection. While some of these equipment may be costly, they're worth the cost when you use them efficiently.

The metal detectors below are great for Gold discovery and I've utilized them to locate high-quality gold:

1. Gold Monster 1000: It works automatically and in an easy-to-use method. It's a high-performance detector that features automatic noise cancel and automatic ground balance, as well as an automatic sensitivity.

2. The GPX-4500 is an Minelab Gold Metal Detector that is able to detect gold in various sizes. It's built with advanced technology and is able to penetrate into

mineralized ground better than most Pulse Induction Metal Detectors. It comes with six search modes that you can modify using your preferred settings and rename them.

3. The X-Terra705 Gold Pack Metal detector
It is a second Gold prospecting detector that makes use of VFLEX technology and an exact pre-set Prospecting Mode. It is the perfect gold detector for beginners. I'm sure you're concerned about what VFLEX is about. VFLEX offers enhanced detecting capabilities with a perfect sine wave transmission as well as the in-coil booster of signal, and frequency selectable for the coil.

Other Gold metal detectors you could purchase to help with your Gold prospecting
4. Whites Goldmaster GMT and
5. Fisher Gold Bug 2
Metal Detectors for Kids
The statement "experience is the most effective educator" has been proven to be accurate and applicable. If you've already taken up the metal detecting and enjoy what you are doing and you want to share your knowledge with the younger generation. If you have kids and you want to show them how to begin prospecting for

treasure even when they're young. If you begin teaching your children about metal detecting as young children it is likely that they will be experts by the time they reach into adulthood.

I've had experience with metal detection since I was a part of the team as in my teens. Within this segment, you'll learn about metal detectors for kids. You can buy any metal detector for your child to teach along with you. With no effort Metal detectors for children can be described as follows:

1. Garrett ACE 300
2. Fisher F4
3. Bounty Hunter Junior T. I. D
4. Bounty Hunter Tracker IV and
5. Fisher F22

Multipurpose Metal Detectors

As a newbie is likely to be enticed to put your hands on every kind of treasure made from metal. For instance, whether it's gold Relics, Silver or coins you'll want the ability to get your hands on anything. If you are feeling this way then you'll require a multi-purpose metal detectors. The configurations of detectors may differ. For some detectors it is necessary to set the machine in All metal mode, so as to recognize any metal it encounters.

On the other side, you are able to pick any metal you wish to search for using the control box on the machine.

Examples of multi-purpose metal detectors include:

1. Fisher 70 multi-purpose metal detector
2. Garrett ACE 150 metal detector
3. Bounty Hunter TK4 Tracker IV metal detector
4. XP Deus Wireless metal detector and more

Chapter 8: The Detection Of Water And The Treasure Identification And Other Instructions

The water that flows through lakes, oceans, and rivers contain a wealth of treasures. The treasures located under water cut across Gold, Silver, and other relics. The lakes, oceans, and rivers contain a wealth of treasures that have been discovered. It's not the case that all metal detectorists is able to hunt in these areas and as a result have a wealth of treasures that aren't discovered to the present.

If you are planning to hunt in the ocean it is essential to know how to swim or dive extremely well. If you aren't able to demonstrate the skills required to be diving, it will first make you unqualified from hunting in that area. Therefore, the amount of metal detectorists who would be able to go out and hunt has decreased. There are many treasures that were lost by travelers many time ago due to their boat crashed. Who will find these items and earn some money out of them in the future?

What are the possibilities for Treasures to be found from Rivers and Oceans?

The probable finds that can be found in River and oceans include:

1. Gold
2. Old coins
3. Silver
4. Relics
5. Rings and
6. Jewelries

In the search for a the river, if you're looking is to locate gold, then you must look at areas in the river with outcrops of gold-bearing quartz rock. There is a good chance to locate such metal in this location more than in any other region in the river. The Gold could be small. However, regardless of its size, it's an important material. You can take a scoop of sand from the river and utilize your detector to look for Gold at the exact location.

Metal Detecting at Beach

I would wish you to find something fascinating within this subheading. The reason I do this is that I am not happy to watch novices at this important exercise suffer too much seeking out treasures at the beach. It is painful digging a large hole in beach sand in the hope that you're getting signals. This signal could be a bad signal due to the environmental conditions the metal detector is located in.

Before you decide to go metal finding metal in a beach there are a few things you should consider first: What kind of water is at the beach I am going to be in to? Are they fresh or salt water? Which metal detector will work best in this water to provide me with the most effective results I require? Is my metal detector made for wet ground or not? These are the issues you should be asking yourself and answering prior to attempting metal detecting at any beach.

I encountered a young man who was digging a gravely into the beach, claiming that he was receiving signals for treasure. When I arrived at the area where he was stressing himself, I watched. I asked him politely why he was digging so deep, and he said he was receiving signals the metal detector buzzing at the spot.

I gently lifted the detector and inspected it. I realized that the detector wasn't designed for this conditions. I smiled inside because I was aware that the young man was receiving bad signals. I explained to the young man that there is nothing to treasure to be found in the area the he was digging. I also explained the reason he had been receiving poor signals due to the fact that this detector wasn't designed to be used on saltwater beaches. I then provided him with the name of a basic metal detector that he could to buy for himself for this type of beach. Metal detectors that are Very Low Frequency (VLF) equipped with ground adjust controls are able to tackle the salt beach detecting challenges. This feature allows you are able to alter the machine into the mineralization level you'd like. At best, however metal detectors that use pulse induction can do the job well.

Certain beaches in saltwater are rich in minerals. Because of it, your detector could be affected if you don't use the correct detector. Take this in mind prior to going to any beach to search for treasures. It is crucial to be aware of that.

Beach Metal Detectors

The following are some sound metal detectors that you can purchase to hunt for treasures on the beach:

1. Whites TDI BeachHunter metal detector
2. White's Spectra V3i metal detector
3. Garrett Sea Hunter Mark II Metal detector that is under the water, and 2 search Coils
4. Bounty Hunter Platinum metal detector
5. Garrett ATX metal detector and
6. Makro Multi-Kruzer metal detector

Locating your Finds using the Metal Detector's Feature

With the latest advancements in technology in the metal detector hobby you will be able to determine what your metal detector is detecting before digging up the treasure. This is a great concept that makes finding treasures a breeze for us. When I first used this type of metallic detector, I was shocked but it's now commonplace as other technologies are advancing.

Utilizing a technique known as target identification indicator, which will inform you of the potential underground find and you can determine the nature of the find the detector is able to detect. It could include Gold, Silver, bronze or even artifacts. This information will be displayed on the LCD of your metal detector's screen. Another method to determine the location is by the signal generated to the detector.

Further explanation of targets ID, every metals are represented by a different targets ID numbers. Additionally, the targets IDs that explain any particular metal discovered differs according to manufacturer of the metal detector. My Teknetics metal detector when it shows a target ID of 88 while I am searching for treasures, it's probably a silver dime. Other metals also come with their own unique numbers.

There is no need to fret about an indicator for the identification of targets in metals displayed on the display of your metal detector. This is because it is described in the instruction manual for the detector you own. Check out the instruction manual for your machine and you'll find the targets indicator codes for each metal that is identified by your device. The introduction of the target recognition indicator on metal detectors has helped us avoid digging for anything that is displayed by the monitor as precious metals when they're not.

Certain metal detectors rather than using the term"target identification indicator use Visual Display Indication (VDI) however all convey the similar message. These are all methods to determine the metal that could be that is found prior to digging. Whites detectors employ"VDI" which is a term used to describe Visual Display Indication.

For those who are new to the metal detectors I'd advise you to start by digging everything. As you become proficient with your detector over time you will be able to start tracking the target ID as well as VDI which includes using discrimination mode. Each step is taken at one time.

How to Identify Your Coins

Coins are treasures that are valuable in treasure hunts. The price of coins can vary. Certain coins are sold at a the highest cost, while others are priced lower. Coins dealers value coins differently. In this section I'll be showing you how to determine the worth of your coins by recognizing them. If you have spotted an item during your the treasure hunt and you want to recognize it by knowing how old the coin is as well as the origin of the coin, its location as well as the value or determine the value of the coin Follow these steps:

* Determine the denomination of the coins.

* Determine the date on the coins along with the shape

Learn about the size of the coins , and take the measurement of their diameters

* Identify the shades of the coins.

Make sure you know the proper words on the coins as it will help you identify the country that could be the source for these coins.

* Download and install the coin checking software on your phone to look up details of the coins. If you own an Android phone you can download and install the application from Playstore. If you use iOS phones, you can download and install the app from App Store. Simply search for coins checker and you'll see a variety of options. If you live in the United States, you can download and install the application known as U.S Coin Checker.

Upload photos of the coins you have found on forums for metal detecting enthusiasts as members may have details about the coins.

Send an email with the image of the coin to dealers in coin. They will be able to identify the treasure and provide you with feedback.

* You can make use of search engines like Google to conduct a search on the images on the coins. Wikipedia might have written an article on the same images. If Wikipedia isn't able to provide an article about it, another blogger could have written something about the subject.

How to recognize Relics

It is possible to find objects that are regarded as culturally significant but you don't know what they really mean. It might appear odd however, it is cherished by many. The identification of the time isn't an easy task. If you want to confirm that the relics are real or artifacts that you have discovered, you must search for archeologists. They will be able to tell you the truth.

How do you identify jewelries?

I am sure that when you look at jewelries such as necklaces, rings, and bracelets, you are able to recognize them. It shouldn't pose a problem for you. The issue is identifying the one that is genuine and authentic. If they're jewelries that bear the hullmark, it is easy to determine them. If they are not it is best to meet the with the jeweler to verify their authenticity.

Accessories Required to use Metal Detecting
If you're looking to become an expert in metal detecting and you want to be successful, then have the proper equipment. Finding the best tools is more than owning the metal detector. It is essential to have backups before you begin your journey. I'll guide you the essential accessories that you must hunt with.
Metal detector Bags
Metal detector bags are essential to detect metal. It's a crucial accessory. When you go on hunts be sure to go with it. The bag will store the treasures that you discover inside as well as other equipment you'll be traveling with.
A Pinpointer

The pinpointer can be described as a hand-held metal-detecting device. It will lead you to the exact location of where the target you're searching for is. When a metal detector can't be reached, a pinpointer will go to the spot. It is shaped like a wand and this is the reason it's possible. It's also known as a probe. A pinpointer can be described as an itch that leads detectorists to pinpoint any object. Certain pinpointers are waterproof, and others do not. They can be purchased at metal detector stores or even on Amazon.

Safety Boots and Apron

Being dirty when digging for treasures is normal , however wearing an apron on your neck can reduce the risk. Apron is a great option to go into the field of metal detecting. Also, carrying an the extra pair of safety boots in your backpack is not an issue. It will help keep your feet protected from any kind of injury if the pair you are wearing gets damaged.

Batteries

If the batteries in your metal detector have gone out of service and you're in a position to know the cause, then you'll realize the importance of going out detecting with additional batteries. Don't take part in any metal detector outing without bringing spare batteries. The batteries in your device will weaken anytime. If the batteries are drained in a setting that doesn't have a shop for you to purchase and replace it right away and you'll be regretting the next day. Be sure to follow this advice: do not shop for Metal Detection without EXTRA BATTERIES. Choose the batteries which are compatible with the metal detector you are using.

Shovel

There isn't a method of metal detector that isn't based on digging holes. Because you will have to dig holes, you have to have a powerful shovel. Furthermore, shovels are an essential tool for hunting relics due to the fact that the majority of the relics submerged in the earth. It is possible to purchase a metal detector shovels from any dealer that is in your area. They're numerous, and you are able to choose the one that's suitable for your needs.

Headphones

Many metal detectors are equipped with headphones. They are an essential part of the modern metal detector. If you are wearing a headphone when you are searching for any metal treasure you will be able to hear the audio clearly. This is different than the speaker on the machine you most likely to miss targets using. The headphone transmits the sound to your ear. In this case you may hear even the most low-pitched tone created by the machine , which could result from the tiny size of the device.

Digger

There is a chance that you will encounter hardly filled sand while you embark on your search for treasure. Due to this, you will require a large digger. It will assist you in making the sand disappear and then you go on to find your prize.

Pouch

This is the instrument you should use to store all your metal detector treasures in. It's a tool that will ensure maximum security for your treasures. There are detectorists who have a particular pouch they carry around however, you can also get yourself any.

How do you recover the Target

If you aren't aware of how to locate the object you are looking for when it is detected by metal detectors and you don't know how, you could make mistakes. It is possible to cause damage to the exterior of your treasure , or even ruin it totally.

To get your target back in good order, you must adhere to these guidelines:

1. If your detector discovers the treasure, when you dig a hole cut a c-shape in the size of a half circle. Then, dig instead of digging all area. When you are digging through the holes, in case an obstruction is blocking the way, get it removed by using a shovel, and continue moving.

2. Find your treasure using your hands after you get close to it. However, if you discover it's still in the distance away, you can use a pinpointer tool to find the exact location to the prize until you reach it.

3. After you have retrieved your treasure make sure you fill the hole by putting the sand that you excavated. Make sure to step over it to ensure that the ground appears neat and tidy.

Doing Good Research prior to Treasure Hunting outing

At a professional level it is not possible to embark on treasure hunting without conducting a thorough study about the location you wish to visit for metal detector. I would like you to begin doing this on time since it will help you become an expert on time.

In this case the term "research" is a reference to the systematic study of the subject. I will guide you how to carry out this research. To conduct your research, steps below:

Find the place you'd like to search

You must determine the area or the environment in which you'd like to visit for to begin your metal detecting. The location could be an abandoned house or an old war zone, or a place that is covered in stones.

Find out more at the library. details about the library.

The truth is that wisdom can be found in books. When you read books with quality learn, you will gain information that can be useful in a couple of areas. Going to the libraries is an excellent way to get a better understanding of the place you're planning to explore.

It is possible to ask the librarian who is who is in charge of sorting books that you want to find an appropriate book for your subject of interest. This will provide you with the facts that are fundamental. When you've finished reading throughout the text, you'll discover the treasures that might be available at that particular location. This will assist you in making a the decision of whether you want to visit the location or not.

It is possible to take note of this and request permission

While you are conducting your research, be sure to make notes so as not to miss any crucial information that could aid in you in your search. Write down the most important details regarding the area. After that, you must get permission from the people who manage the environment before continuing in your hunt.

Find the primary zone of the area you want to target.

After you've gotten details about the area at the library you are now able to identify the area that you wish to find on. Use the maps of the area to complete the task. When you're done you will be at the primary area in you "treasures collection".

Chapter 9: Increasing Your Detecting Skills
Clean And Selling Your Finds

In the course of our lives, we develop in our own individual abilities by working in the field of expertise. The same is true for metal detecting hobbies. You will improve your skills as you learn and practice. You also discover something new on the job.

This chapter I'll explain what you can take to enhance your abilities. It is important to take lessons from experts in order to improve your skills. If you're interested in seeing further than you can as a person one method that you can employ is standing on the shoulders of an enormous. In this sport there are giants who you can place on their shoulders and see in a wide area. They are usually found in metal detecting clubs and communities.

Once you have discovered your treasures with your metal detector The next thing to think about is how to sell and clean your treasures. This will not be an problem since I will guide you in this section. I will show you the best way to sell the treasures you discover.

Joining Metal Detecting Club

A lot of beginners don't realize what one of the most important ways to increase their skills in detecting, aside from continuous practice is by joining metal detector clubs. There are a variety of metal detecting groups that can assist you in growing. I'll list them and invite you to join one or more of them to help you improve your skills within the realm of treasure hunting.

The metal-detecting clubs work in the following order:

It is the Warrior Basin Treasure Hunters Association (WBTHA)

The metal detecting club is situated in the Birmingham, Alabama area, in the United States. The goal of the group is for members to support each other in sharing an fascination with metal detect. The types of metals they concentrate in include the detecting of coins, relics gold, rocks, and more. They also have meetings. You can become an individual member. The club was established in 1972.

Central Alabama Artifacts Society

Another metal-detecting club you can join with no pressure. They assist each other with developing their skills. This club can be found at Alabama, United States. The club usually meets on the last Tuesdays of each month.

The Sal's Historical Hunt Club

The Sal's History Hunt or simply put History Hunt Club is a metal detecting club for hobbyists. It was started in the year 2000 by Sal Guttuso. Every year, club members travel to treasure hunting across different countries , primarily Americas in addition to Europe. They suggest joining their Facebook group under"History Hunts Metal Detecting Group "History Hunts Metal Detecting group" prior to signing up to join the club. On their page on Facebook, members will be able to get to know other detectorists with experience.

South East Treasure Hunters

There is another metal detector club that you can join to increase the metal detector skills you have. It's located at "789 Brentwood Drive Gadsden, AL 35901. Members assist each other in identifying the development of their skills.

If you live in the United Kingdom, there are several UK metal detector clubs you could join. Some of them include:

1. Central Searchers Metal Detecting Club
2. Wessex Metal Detecting Association
3. The Magiovinium Metal Detecting Club
4. Cleveland Discoverers and
5. Plymouth Detector Club

How do you remove Your Found Coins

I will help you learn how to clean your coins made by Silver or Copper. It is essential to follow this procedure to ensure the coins as tidy as they can be.

The steps you need to take to cleanse your coins:

1. Cleanse your hands thoroughly using soap and water prior to begin cleaning your coins

2. Then, open a faucet of water and let the water to pass through the coins and knock off dirt from the surface of the coins. The water should flow with greater intensity.

3. Warm water boil and then pour the water into the bowl. Add detergent and mix thoroughly.

4. Incorporate your coins into the mixture , and then shake the bowl that holds the coins and the mix slowly. There are a few coins required to add.

5. Then, wash the coins in distillate water

6. Dry the coins and clean the surface lightly with a an absorbent towel

Other strategies you can employ to wash your coin include:

1. Use of warm flowing water

2. Utilizing a toothpick or soft bristled toothbrush to clean the coin's dirt and wash them using a mix of soap and hot water then let them air dry.

Take note: Do not scratch or scrape the surfaces of coins as a result of cleaning as it can cause damage to the coins. When it happens, nobody would want to purchase them.

Step-by-Step Guide to Cleaning your Jewelry that has been uncovered

To cleanse your jewelry you spotted during the treasure hunt, follow these steps:

1. Make sure you wash your hands correctly using the use of soap that is clean and with water.

2. Mix a few drops liquid dish soap and cups of water

3. Put the jewelry into the mixture and let it sit for a couple of minutes

4. Eliminate them from the mix and then wash the mixture with clean , distilled water

5. Clean them with a soft towel. them and then let them air dry.

How to clean up your gold Find

Gold is a unique metal. Due to the precious nature of this metal can be, it will not contain a lot of dirt. A gold is always Gold.

To wash Gold simply use warm water and add a small amount of liquid soap. Incubate a soft towel in the mixture , and then use it to scrub the Gold. Afterthat, allow it to dry in the air. Do not employ ammonia or bleach to clean gold due to their roughness.

Cleaning Iron Relics by Electrolysis Method

Electrolysis has been proven to be the best method to get rid of Iron Relics back to a good condition. The word "electrolysis" refers to the act that uses a direct electrical current to stimulate an otherwise inexplicably chemical process". Because Iron Relics are prone to corroding it is imperative to use electrolysis to clean the area becomes necessary.

Electrolytic kits aren't that expensive and you can purchase a kit to clean your own process. You can make one yourself should you be skilled. You can carry out this purification procedure in a cool area. The reason for this is that hydrogen gas released in this process isn't harmful to the human body.

The steps for cleaning your Iron Relics with electrolysis are as follows:

1. Clean the Iron Relic with water and soap, and ensure it's free of sand and oil.

2. Put some water into an unbreakable bowl, and then put baking soda in it.

1. Let the baking soda completely dissolve in the water. The baking soda that you have added will be charged by the water. Stir it until it completely dissolves

2. Attach the positive terminal of the Iron Relic that you would like to wash

3. Attach the negative terminal of the metal of sacrifice (the metal that will sacrificed). The metal could be made from Steel. The electrolysis process continues it causes rust to form on the Iron Relics is transferred to the sacrificial iron and is referred to as the anode

4. Turn on the electrolytic cell by using the speed charger.

5. Let the cleaning process continue

6. The process should continue for at minimum 8 hours, or until the rust on the

Iron Relic is gone. It is possible to set up the process at late night and continue it until sunrise.

7. If, at some point, the anode metal is completely covered by the rust of the relic, shut off the setup to replace the sacrifice material with another metal of the same type (in this case Steel)

8. After that, switch off the electrolytic set-up take out the cleaned Iron Relic, rinse it with fresh water and dry it.

How to Preserve Your Iron Relics

You might want to save your Iron items until you can contact an interested buyer. For this it is possible to use the same method I have used to preserve my personal Iron Relics:

1. Dry your relics in the an oven at about 200 degrees Celsius

2. Then, remove the relics with thongs . Then pour the oil from paraffin onto it since it's still hot. Don't touch it with your hands in case you get injured since the metal is hot while it is hot.

3. Let the oil dry in the Relics

4. Use a soft brush to sweep the surface

If you are in this state you've preserved your Iron Relics.

How to Sell Your Finds

While many detectorists don't want selling their finds, there are some who sell their finds. You can offer different kinds of items to various people.

1. You can also sell your finds to shops. You can offer your found jewellery, gold, coins and Silver to various shops on the internet or offline. Look around in your neighborhood.

2. You can offer any old coin to a coin dealer. The amount they'll buy is contingent on the value they place in the old coins.

3. You can offer any old coin to sites that buy them.

4. You can also sell your old coins and artifacts to museums that require the items.

5. Dealers in Silver and Gold are always available to purchase these precious metals from you.

Chapter 10: Getting The Language Of Metal Detection

The art of learning the language of metal detection is a great experience, and I'll be teaching you a few phrases to recognize metal. It could be useful for you prior to going out to find the most precious metals.

All metals: The majority of metals come with discrimination targets. If you detect the target without the discrimination signal switched on, will cause all metal objects to generate a signals. The discrimination feature allows you to eliminate some undesirable objects. All metal mode is typically utilized in a clean site where only precious metals will be found.

Bling is extravagant jewelry that could be made of precious metal, or not be a precious metal. It's a wonderful opportunity to discover Bling. The bling can make you feel amazed at first glance. The most beautiful bling is piece of jewelry that is stuffed with precious metal

Tone ID can be described as an audio message from a metal detectors that inform that you are in the vicinity of treasures prior to you begin digging.

Black dirt: It's naturally rich common soil that is found in very old locations. The appearance of black dirt may be an indication that you're closer to the treasure. The technology of visual identification has come further. The producers of detectors have also created metal detectors that come with a tiny screen that can identify what's hiding under the soil before you begin digging.

Black Sand: Black sand is an excellent indicator that you're nearer to gold. It's like the iron particles which are tiny. It looks like sand but they're not. This is a good thing at gold-hunting locations however not at the old coin-hunting sites. You must adjust the metal detector you use as in certain situations the detector may be unable to working in these locations. Some detectors might not function at all. This can cause havoc for some detectors made of metal.

Bucketlister It's a exclusive one-of-a-kind discover.

Cache: These are jewellery or coins that are intentionally buried together by a person or group of people for a lengthy period. They're usually placed in containers, jars, or other vessel. The term "cache" can be described as an "cluster" of coins or valuable objects that are found in close proximity but not necessarily in the same location.

Cache hunting is specifically looking for clusters of precious objects or coins that are found close to one another. It is a different method of locating an area than normal metal detectors. If treasures were found in areas where the conditions were met, such as close to a spot in which animals live, or near landmarks which could be found easily. The hunt for metal is effortless in these areas.

Canslaw The Canslaw is a set of aluminum cans that were left over after having been hit by a lawnmowers. It appears like people deliberately scattered the cans. These cans provide a range of signals because of their dimensions and cause a difficult search in the surrounding.

Color The term "color" is that refers to gold due to its color.

Choppy is the sound the sound that a detector emits when it detects something that can be separated from. It is usually used to define a suspect signal. Some coins do have choppy signal.

The Clad coins are new coins that have been made with mostly non-precious materials. In the USA these are usually silver-colored coins. They are typically an indication of contemporary activities. If they're not present in the hunt, but older coins are in the area this is a very desirable thing at the spot.

Coil: They are called loop in certain countries. It's a round wire that is on the other end of the metal detector.

Coilball is an element of dirt that has an actual coin in it. They're always great. They are often the final moment of anticipation before you discover the kind of coin and the age. They are healthier for the environment. Sometimes you will notice that the edge is Copper.

Coin spills happen when a coil falls out of a bag filled with detectorists, without notification due to the tiny dimensions of the bags.

Digger: It is a tool that is used to locate the targets. It is also described as the person who can detect. "Hello, diggers!"

Pennyweight: It's an amount of weight equivalent at 24 grams.

Friends The term "friends" refers to groups of individuals who have great targets within a single hole. The hole was fought by his group of friends.

Pinpointer: Small hand-held metal detector, that fits within the plug or hole to aid in locating the object.

Assay: This is the method you can use to determine the purity of your treasure, whether it is silver, gold or any other type of metal.

Tot Tot The name may seem odd to you however it's actually meant to symbolize the playground. It's actually a good and enjoyable to explore.

Pinpointing is the act of reducing the target area to an area that is small enough you can dig using the coil that is the main detector or a handheld probe. Metal detectors typically have an "pinpointing mode" that lets users make an audible "X" in the ground's surface.

Plug: It's an excavation that is carefully dug into the ground to ensure that grass and dirt aren't damaged. Digging a quality "plug" will be the sign of a seasoned or ethical detectorist since it minimizes the impact on the properties that are being searched.

BFO stands for Beat Frequency oscillation. These detectors are older and utilize the induction balance principle. It is often found in low-cost metal detectors, and seldom employed in coin shooting. They are often used in conjunction with "old old school" detectorists who do not want to let go of their device.

The Relic Hunters are a person who seeks out common objects, not only precious metals. The hunts take place in forests or fields and frequently targets are a reflection of conflict from the beginning, like those of the Civil War in the US.

How to Use Your Metal Detector for Hunting
If you are using your metal detector in any location you choose the manner in which you set up your metal detector will determine whether you'll find valuable objects or not. A lot of people carry their metal detectors but not properly and often find themselves without treasures for the day. However, those who are experienced in the field will be able to pass through the

same location and uncover treasures. I will show you how to make use of your detector while out in the field to get amazing results.

If you're out on the ground or another location you prefer using a metal detectors, ensure you keep your searchcoil's in the range of 1 to 2 inches. Your detector should be in line with the ground to get results.

Don't work more quickly or walk around like you're in a rush. Remember that you're on the lookout for something crucial, which is why you need to be patient when searching for it. Therefore, you should stroll slowly while scanning your searchcoil from side to side , moving your coil at a rate of around two to five inches per second. Be sure to advance the searchcoil approximately one-half the size of the searchcoil when you finish each scan.

Filling Holes

In the metal-detecting field there is a standard requirement that you must meet, that is filling in the hole. A lot of people make this common error in their hunts. The act of not filling in the hole that you dig when you are removing treasure from beneath is extremely dangerous. If you don't complete the filling, the area appears like a war zone, and the erosion effects to be felt throughout the surroundings. Don't leave the site you have dug without properly covering the hole since it is essential.

Property destruction

When you hunt on any area you have been given the right to hunt you must ensure that you do not ruin any property you're trying to retrieve treasure. The types of properties that you should not be destroying destruction of are trees that kill animals, or destruction of plants and trees. You should only take away only what you are allowed to take away.

Take out trash

When you dig to find treasure, you can be 70% chance that you'll take out rubbish. It's good to find garbage. Keep in mind that you're digging your way through the earth, and a the trash you encounter is something that happens. When you find garbage, take them out. Following your discovery of the treasure that you found earlier, you can take the trash along with you and put the items in a garbage bin. This is a pleasant way to spend your time.

Monitor the environment

If you are in a spot to hunt, don't be in a rush to hunt, and then leave. Do your best to take your time and observe the surroundings. For instance, if you go to the area through an entrance and the gate is shut at the time you enter, attempt to shut it down after you have left. If it was opened when you entered, you can try and open it when you leave. If you have anything that you took out while excavating your treasure Try to place them back exactly in the spot where you first encountered with them, so that the owner won't be upset with the action.

Chapter 11: The Things Your Detector Is Telling You

Technologists are a blessing because every day they are developing metal detectors with added features. The detectors have been modernized. They now offer information that allows you know what you've found prior to digging them. If you're interested in and would like to dig to know more about, you may take them out. However, If they're not your style you could leave them as is instead of spending your time in them. In this section, I'll teach you how to read the detector will display to make it easier for you to understand.

Audio Tone

When you use a metal detector, an audio tone is generated when the detector is able to detect an object. A nonferrous target detected will provide a medium - loud audio sound, whereas an identified ferrous target will provide a low-tone audio response. Tone is produced according to the type of target or treasure it detects. Make sure to pay attention to the detector tone.

Target Indicator

Metal detectors have targets that are often referred as treasures. The indicator for targets indicates which treasure is hidden within the coils.

Target ID Number

The numbers show the number of the targets found. The numbers do vary in variations based according to the maker. Some range from -4 to 44 for the X-Terra 305, and -9 to 48 for the X-Terra 505. Negative numbers are ferrous targets while positive numbers indicate
Nonferrous objects.

Depth Indicator

The depth indicator provides an indicator of how deep a target can be. It is based off the machine. It informs you of how depth the detector is able to go. This is a reference to shallow, medium or even how deep it will go.

Search Coil

Search coils of metal detectors are among the most important components that make up your detector. Without them, your metal detectors are useless. You might be asking yourself what exactly is a the search coil. The search coil can be described as the piece of wire that is located at the bottom of the metal detector that is used in the detection of metals. They typically come in various dimensions and shapes. Each size has its unique form of strength and weak points. If the coils are considered as a pair, the smaller coil is always superior to the bigger one in the sense of locating treasures. The 8-inch coil is an all-purpose coil that is the most well-known coil for metal detectors. The larger coils are appropriate for areas with low mineral content and in areas where there isn't too much trash. A smaller coil is great for collecting treasures in areas where there is plenty of trash. Larger coils are generally heavier than the smaller ones, and are able to cover a greater the ground. A coil with more than one size is a good idea especially when you have different ground levels, however most detectorists prefer smaller ones.

Know Your Machine

If you're looking to get into the metal detecting passion, you need to understand your equipment. Once you've purchased the detector of your choice first step is to take the time to study the instruction manual thoroughly. Learn use the device. It is possible to test it within your compound, for instance digging up a gold chain or any other metal that you want to employ an instrument to identify the metals. This will help in understanding the way your metal detector functions. As time passes, different firms will release different types of equipment. There is no need to purchase a different type of machines every year. Learn to operate the machine you already have. If you can master them hunting with them can be fun.

Making time to master and drinking your machine is definitely time well-spent. It is also possible to create a small-scale hunts and then bury diverse treasures on various locations. They'll help in helping you understand the different ways to play.

Mineralization

Let's discuss the effects of mineralization and how it affects the metal detector when you go on hunts. Mineral is a natural chemical. This means that almost all areas of the earth contain natural minerals. There are various kinds of minerals, however, we will examine the most prevalent one that is used in metal detection like nuggets gold. One mineral type is deep mineralization of the ground, where the ground is hot by itself and may contain hot rocks. The another form of mineral is the moderate one which has little mineralization of the ground.

A few of the naturally occurring minerals such as salt, irons, and hematite could impact your metal detector's ability to digging deep into the earth to find treasure as well as cause problems for your detector due to it not being able to distinguish target. You've now learned that the earth has mineral deposits, you may be pondering you're thinking about, which metal detectors can handle these minerals?

If you're looking to find the presence of metal in mineralized ground, the ground metal balance detector a great option since the metal detectors that are equipped with ground balance are ideal for ground that is mineralized. Ground balance can be a method to connect your detector to any kind of environment. When you use a detector with a ground balance is telling your detector to not ignore the earth minerals and instead find the best metal, such as gold, among others.

Metal Detector Classes

Metal detectors are simple to use, there's an electronic instrument that is precise. Understanding the process of technology fundamentals will go a long way toward knowing how to set your detector's control in various conditions and get the most value out of it.

White's Metal Detector

White's metal detector isn't only a brand-new detector. You may be experiencing this name only for the very first time. This detector has been on market for quite a long period and comes with a range of brands that you can choose. It is one of my favorite brands of metal detectors. This detector is less expensive. For those who are just beginning to learn about metal detectors, you could opt for this kind of metal detector.

They're of high-quality robust and durable. They are also user-friendly. I can remember clearly in the year 2016 , when I bought the metal detector. I saved some money before

I could come up with the exact price to purchase it. I am very happy I own a metal detector because it's been a great help in finding several coins as well as other relics.

For a number of years the metal detector has served me well, and I keep it in my home. The detectors are friendly and can detect valuable items like coins, gold, relics jewellery, and more.

Garret ACE 250 Metal Detector

Fig 3: Garrett ACE 250 metal detector
Garrett ACE 250 is the most efficient metal detector available for cost. Its weight is 2.7bls. This makes it lightweight and easy to transport. Garrett ACE 250 has a waterproof characteristic, which means it is not damaged by water in shallow depths. It can identify metals like jewelry, coins and silver, gold as well as relics .Garrett metal detector is compatible with headphones however it is recommended to purchase an upgrade because the one included with Garrett Ace 250 does not have a higher volume. Therefore, you should choose a model with

a higher volume. It is the Garrett ACE 250 metal detector is often referred to as a fantastic beginner's metal detector. It is durable light weight, simple to use , and extremely affordable. It's an excellent model to start with if you want to try your hand at the metal detecting as a hobby.

You can go hunting with your Garrett ACE 250 Metal Detector

If you are hunting with a metal detector like the Garrett ACE 250 and you are just beginning your journey on the scene, I would recommend that you start your hunt in a sandy space like playgrounds or on beaches. They will help you hunt easier as a newbie and will assist you in learning the operation of your detector. Garrett ACE 250 is really an excellent metal detector once you understand how to use it properly.

Can Garrett ACE 250 Find Gold?

Yes, Garrett ACE 250 can find gold. Also that all metal detectors are able to detect gold. It's all about the depth of the gold and how big it is. The bigger the nugget, the more difficult it is to find. The ultimate response to that question would be the process of finding gold is dependent on the surface. It could contain iron minerals as well as salt or both. The more you have salt or mineral iron the more difficult it becomes to find a hidden object.

Are a Garrett ACE 250 a Good all-rounder?

Yes, Garrett 250 metal detector is an excellent all-purpose detector and is a breeze to get started. Be aware when you are using it on the ocean since it is only water-proof, while the control panel isn't. If you do not plan to use it in the water, it's great all-around.

Teknetics Metal Detector

This metal detector is as efficient as a bounty hunter TK4 tracker IV. The Teknetics detectors have a variety of options that let them be utilized in a variety of places. The detector is owned and operated by the very first metal detecting firm around the globe, Fisher Metal Detector.

It's a amazing metal detectors that which you can use to search for treasures at various locations you prefer without anxiety. The process of learning about how they operate is simple. It is not necessary to spend all day trying to find someone who can teach you how to use it. The detector is

easy to master. It also comes with a reasonable price label. The majority of Teknetics products are sold for sale at a fair cost. It is not necessary to pay hundreds of dollars for an upgraded detector. In fact, it's a great choice for novices to begin with.
Minelab Metal Detector

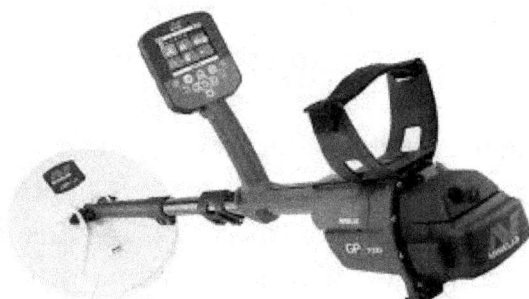

Fig 3.1: Minelab GZP 7000
A metal detector from Minelab has the capability of detecting gold. It is a gold prospecting model. Prospecting models for gold can be costly. Metal detecting can be a fascinating and rewarding hobby that is enjoyed by a lot of people all across the globe, with an thrill of discovering precious objects like jewelry, silver, gold and more.

In this article discussion, we'll look at Minelab GPA 7000 as it is the most recent gold detector. However, they also have smaller ones that are also good. GPZ 7700 is a very effective metal detector for hunting gold.

It's also a top selling metal detector on the market. Minelab is a top-selling metal detector. Minelab metal detector is capable of handling nearly every kind of situation. It is able to differentiate between different types of target (treasure) types , and you can program them to block or eliminate undesirable particulates (trash).

It is a detector made of metal to meet the professional standards of the detection of gold. The GPZ 7700 is the top gold detector available on the market. This detector is more sensitive than the normal value, which is not see on other detectors made of metal. It has a high sensitiveness and depth for gold. The distinction between GPZ the 7000 in comparison to other detectors is it's constructed using the technology of zero voltage transmission.

This device can detect gold as much as 40% more deeply than other detectors since the majority of gold is found hard environments that have a lot of mineralization, but the GPZ 7700 makes finding gold in these areas easy thanks to its accurate automatic detection ground balance. GPZ 700 is pricey. However, for serious gold hunters this is the best option for these. There's the Minelab GPZ 5500, which is less expensive should you be unable to be able to afford GPZ 7.7000.

Frequently asked questions

Can the Minelab detectors detect all types of gold?

Yes. Minelab offers a selection of gold detectors that cover all prospecting levels, including professionals who specialize in gold prospecting small-scale mining by hand and serious gold prospecting for the holiday season as well as weekend enthusiasts and the collection of gold samples. It is able to detect all forms of gold.

Can a Minelab metal detector tell me what an object is likely to be prior to me mine it?

Yes. Minelab metal detectors have the capability to distinguish between various types of treasures. The discrimination function on Minelab detectors measures two important properties , which are ferrous and conductive. It detects garbage based on their conductivity properties.

XP Metal Detector

XP Metal detectors are among the most prominent producers of metal detectors in the world. The detectors produced by this company are extremely light and compact. They are extremely comfortable as well as speed and performance. Metal detectors are constructed precisely.

It has a high-quality performance that has quickly become a standard among enthusiasts. It's great at discovering dimension quickly deep, light, and completely wireless. XP metal detector comes with Wireless Remote-Control Display Screen Backhead Headphone with an electronic control that can be removed. There's no wire which could disturb you while you are out hunting. There are a myriad of testimonials regarding this metal detector from various perspectives.

It is a good option as it can detect hidden treasures that most metal detectors may not be able to find. The metal detector is ideal for those who love to hunt. If you're a novice, this is not the ideal option for you as it's mostly experienced metal hunters who can manage this type of detector. This detector is able to work well in wet beaches dry beach, dry and relics, gold, coins hunting, and more.

Bounty Hunter TK4 Tracker IV

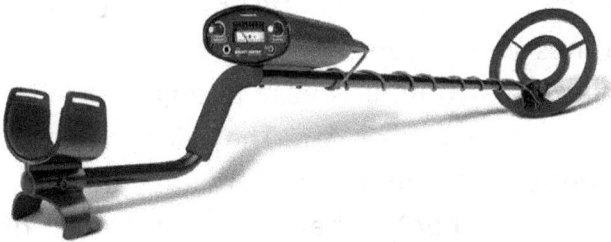

Fig 3.2 Bounty Hunter tk4 Tracker Metal Detector

Bounty Hunter TK4 Tracker IV is a basic and cost-effective metal detector that is perfect for people who are looking for an easy out of the bag metal detectors. It's also a great tool for children to train the youngsters on how to discover treasures that are hidden within the compound or in the backyard of the home. The Bounty Hunter TK4 Tracker IV detector is ideal for people who have never tried a metal detectors before and wish to start at a lower cost. It is a low-cost metal detector priced at around $130. It could be even less than the price. It's a simple detector with an analog display. It features sensitivity and discrimination adjustments, as well as the motion of all metal mood and two tones of audio-based discrimination. Bounty Hunter TK4 tracker IV is an 6.7khz VLF (VLF) detector that comes with three operating modes and an adjustable stem. It is easy to adjust the size to the size you prefer. They designed this product to be durable to withstand any kind of environment, whether it be tough or smooth. It is a motion metal detector. It has to be moving to be able to be able to identify metal.

There is a built-in speaker with a the 1/4 headphone Jack however I would suggest buying a headphone in order that you don't lose a focus. A headphone jack purchase is likely to be an excellent choice to ensure that you are in a position to block out any background noise so as not to be distracted by any kind of sound which could be coming from any part of the world. The bounty hunter's TK4 tracker II metal detector is well-organized making it user-friendly and also easy to understand. With just two knobs , and two tone audio, the device offers an easy learning curve however, they are able to handle certain threatening circumstances. The analog meter measures signal strength that shows size and depth of the target. The weight of the bounty hunter isn't too heavy as it weighs 4.2lbs. It's not the tiniest detector, however, children are also able to benefit from it since it is not too heavy to transport. They are also able to be used for hunting for long durations. There is no adjustment for the ground balance function, and as such it is not able to work on mineral-rich ground.

Settings and Features

Figure 3.3 Control panel of the bounty hunter TK4 Tracker IV metal detector

Hunter of bounty TK4 tracker IV detector made of metal comes with three settings that can be adjusted, which include the sensitivity, search mode, and discrimination. We'll explain the role of each of these adjustments.

Sensitivity

In the bounty hunter's the TK4 Tracker IV metal detector, the sensitivity does a fantastic job in the hunt for metals. The greater the sensitivity the more accurate the detector's ability to recognize a target however, the lower the sensitivity, the lower the capability that the detection device has to identify the target. However, there's bound to be an issue of greater sensitive. The more depth the detector is able to search, it is more likely to be activated by mineralization of the soil. This is why it is possible to set it to a lower sensitivity to prevent unintentional chatter. You can alter the sensitivity through the control dial to left-hand side of the control panel.

Search mode

Search mode is activated by pressing the switch located at the lower right corner on the panel control. Three search options that are as follows:

1. All metal Mode: All metal is used to identify all kinds of metals, including the gold Silver and aluminum jewelry, copper. They can also be used to hunt for relics.

2. Tone mode has two-tone audio that automatically rejects iron. It also blocks discrimination settings. It emits a either a low or high audio signal that is based on the type of metal.

3. The full discrimination feature makes use of only one tone. Most of the garbage is instantly removed when you use this method. It is possible to adjust the degree of filtering by using the discrimination dial.

Discrimination

Metal detectors that distinguish between metals have the ability of metal detectors to avoid targets it does not want to. This includes metals which is not as clean as lead, and other such. This bounty hunter TK4 IV comes with only one knob for discrimination. The higher the level is, the more forceful the discrimination .But it is also important to be cautious about identifying objects that you are trying to find, as the higher level of discrimination allows you to locate highly conductive metal that does not have an iron signals, such as silver and copper. It might reject certain crucial metals, like brass or gold, and brass.

4. Search coil that is waterproof The stock set includes an 8-foot search coil that is waterproof however, the control box doesn't come with waterproof, so ensure you check that your box will not comply with water.

Many people ask questions

Is bounty hunter TK4 Tracker IV metal detector suitable for children?

It's appropriate for teenagers and children. I personally utilized it at the teenage years and loved it since it can detect every kind of metal. It's also much less expensive.

Does it work to hunt for gold?

As I mentioned earlier the bounty hunter's TK4 tracker detectors can be used to identify all kinds of metals such as copper, silver, gold and many other treasures, but there's one small issue with the gold hunt. The issue is the gold nuggets are usually located in large mineralization zone. Because the detector isn't equipped with automatic adjustment for ground balance the detector will have to contend with lots of chatter within those areas. Therefore, it's not the best option to go for gold.

Do you have a warranty?

Yes, it is. It is a bounty hunter tracker II metal detector comes with a five-year warranty.

Chapter 12: Relics Metal Detector

Relics hunting can be extremely enjoyable as there are numerous types of relics waiting to be found. Utilizing a high-quality metal detector is among the most effective methods to locate many hidden treasures. It will be very lucky in the event that you come across several relics while you are out hunting pursuits. For relics to be found in the first place, you'll need to be in an area that is rich in history. You also need to be selective about what is it that you are looking to find so that you are able to conduct a lot of research on the region before you begin your journey of looking. For instance, if you are looking to explore the past of Spanish time or modern-day hunting at beaches, you have to purchase a metal detectors that are suitable for this type of locations. Hunting for relics is addicting and exciting. You don't need the one of the most expensive detectors in the market prior to hunting for objects. If you're not experienced in metal detectors I suggest that you go with lower priced ones to move up to better ones later on. There are a few metal detectors that are ideal for hunting relics without anxiety.

They include Garrett AT Pro, minelab CTX 3030, Fisher F22 and so on. If you're looking to identify Relics, it is best to choose one of these. They are specifically designed to detect the relics as they are made with an extremely low frequency, in order to pick up specific targets, such as brass, iron and so on. Many of the relics that have been discovered originate from mineralized regions. This is the reason that the metal detectors are not able to detect relics since the majority of metals are difficult to discern in these environments.

It is possible to use a gold metal detectors to find ancient relics. However, not all relics can be located using metal detectors for gold or beach since some relics are made of brass, irons, and steel which metal detectors generally consider to be trash. But some modern detectors feature a relics hunt mode that makes all metals in the collection not classified.

Coins Shooting Metal Detector

The most popular reasons people use their metal detectors to search for coins. If you are searching for coins as your pastime, it's often referred to as coin shooting. Coins shooting is extremely enjoyable and a great method of having fun and putting money into your pockets. I frequently say that if you're a metal detectorist that is serious about hunting and frequently go out looking to find treasure, then you may be able to earn a decent amount of cash out of all the finds you find when you decide to sell your finds. There are many methods I'll teach you that will aid you with your coin shooting passion. If you are interested in coin shooting as a passion, you should follow these guidelines:

* If you're interested in coin shooting, ensure you are familiar with the dealers and get more information about coins as there are various types of coins, and are not able to determine the worth. Inform them about any coins that you can identify to help them understand which one to purchase from you.

* Do not scratch the coins or scrub them too often when you remove them since you may damage precious coins because of the regular washing of their surface. It is

recommended that you give the coins you take out a moderate amount of cleaning or leave them in a moderately crude condition in order to avoid damaging it. It is possible to clean silver since they're in excellent condition. I would suggest cleaning your coins by using water that is laced with soap. When digging for coins, be cautious to ensure that you don't ruin the coins by digging them with the digger.

* If you dig an area to collect coins, do not be in a rush to go away because there could be additional coins around the spot. Make sure to check before you leave.

A pinpointer can be amazing because most coins are tiny and are difficult to locate in a pile of soil that is loose. A Pinpointer by Garrett pro is an excellent alternative.

* If you're an expert at coin shooting you can tell other objects and accept coins as coin shooting is about making the most of your time to do efficient looking. The process of digging up lots of rubbish is not a good idea in the event that you already know what you're searching for.

So, for good, effective, inexpensive coin shooting, I advise you go for Garrett ACE metal detector. Coin shooting machines is also able to identify metals such as irons.

Metal Detector Underwater Detector

Imagine how exciting it would be to discover treasure underwater using the metal detector. The prospect of detecting metal in oceans and beaches is an awesome idea. Technology has advanced to the point that there are metal detectors that is able to withstand the pressure of deep waters. They are driven by electricity. The underwater detector will immediately alert you to the metal object that is below the water surface for you to search no matter if you choose to invest in pulse induction or a very minimal frequency metal detector. Here is the listing of metal detectors can be used in the water. The following are the ones:

* Bounty Hunter Tk4 Tracker underwater metal detector: This gadget assists in detecting metals beneath the water.

* Garrett at prounder water metal detector

* Minelab Equinox 600-multi IQ underground metal detector This detector is a submersible and provides you with maximum output with minimal effort. The

multi-frequency machine can go up to 10 feet below the surface to locate important antique pieces that nobody had ever seen before, be it an ocean, stream or lake, and so on.

* Kuman automatic underground metal detector

* Kkmoon underwater metal detector that lets you discover the treasures you are looking for in the depth of 30 meters. It weighs approximately 1.2 pounds. It features pulse induction that means that the device is appropriate for fresh as well as salt water. It is also not a bulky device. Therefore, don't worry about the transportation.

* RM RICOMAX underwater metal detector. It detects metal underwater that has an extended search coil. It allows you to find deep hidden treasures with the least effort.

Gold Metal Detector

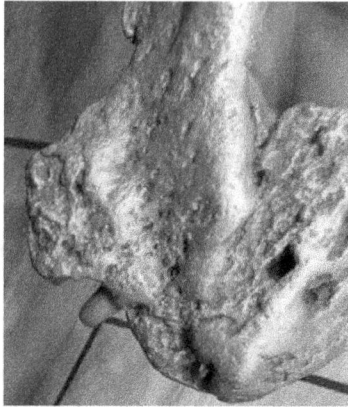

Figure 4: Illustration of a gold nugget

Metal detectors specifically designed for gold is actually a costly one since not all detectors are able to detect gold nuggets. There are specific metal detectors specifically designed for this because if you choose any metal detector without knowing the objects it is specifically designed to locate, you may find no gold in when you finish your hunt. The hunt for gold is an enjoyable activity, therefore don't let it deter you.

Therefore, choosing a metal detector that has been designed to specialize on gold can be a beneficial step because the majority of gold can be located in mineralized areas and not every detector is able to be used in such areas. Here are some of the metal detectors suitable for gold hunting and provide more success in the detection of gold in mineralized as well as non-mineralized regions. But before we dive deep into the matter, I would suggest a factors to keep in mind prior to buying a detector:

Be aware of your budget, as gold detectors are expensive.

Make sure the detector can deal with mineralized zones, which is a crucial factor because there are a few detectors that can handle this.

* Prior to beginning your hunt, ensure you do some research about the place. This is an essential aspect.

* Purchase a detector that has a properly-manual ground balance

* Select a very low frequency metal detectors or pulse induction technology.

These are machines great for gold hunting.

* Minelab GPZ 7000

* Minelab GPZ 5000

* Goldmaster whites GMT

"Fischer gold bug" 2

* Tesoro lobo superTRAQ
* XP DEUS

Metal Detectors for Kids

The purchase of a metal detector to your children is always a good idea. If your children have an passion for metal detecting and you want to encourage them, don't let them put them off. It is possible to help them realize their own goals. My father bought my a detector made of metal when I was my teenage years and I am grateful the purchase was made by him. In the end, I've got excellent experience in the field of metal detectors are and has also helped me learn about things I didn't know existed before. After I went on my hunt, I decided to conduct research on my find and now know many things I was unaware of prior to. Metal detecting can be a great sport for kids. It's educational, sparks interest, and helps children to spend more time outdoors. It's an excellent opportunity for children and adults to be together.

Metal detectors can also assist children and adults learn about the local history. Selecting a metal detector for children isn't an easy task since you don't know what is the most lightweight design, long-lasting and easy to use. It is affordable, and you can purchase it on Amazon to get a quick and easy delivery. Metal detectors for children will be listed as follows:

1. Bounty Hunter Junior T. I. D
2. Bounty Hunter Tracker IV
3. Fisher F22
4. Garrett ACE 200 and
5. Teknetics Digitek Youth Detector Tweens and teens

There are a lot of other models in the market. It is possible to do your own study and find one for your children. The metal detectors I've mentioned have all the necessary features to be lightweight sturdy, reliable, easy to use and priced.

Multipurpose Metal Detector

The multi-purpose metal detector is one of the detectors that is used to identify objects from different locations without having to reject valuable treasures. It can identify coins, silver, gold or relics. In the marketplace there are metal detectors which can do multiple tasks simultaneously. This means it is able to detect various metals. My first metal detector that I worked with was the bounty hunter TK4 tracker Iv , which is a multi-purpose detector. it allowed me to find various kinds of treasures that comprise jewelry, coins and also work very well in beaches.

Multipurpose metal detectors are great for those who are new to the hobby because it can discover all kinds of treasures, both for dry and wet soil before you are able to find treasures you are keen on. When you first start getting into the metal detecting world it is possible that you be enticed to find every treasure. If you are into this category and are looking for a multi-purpose metal detector, this is the best choice to meet your needs.

We have a variety of detectors that can detect diverse treasures. They include Bounty Hunter TK4 Tracker IV Metal Detector, XP Deus Wireless Metal Detector, Garrett ACE 400 Fisher F22 and many more.

Chapter 13: How To Clean Up The Metal Detectors Finds

As a metal-detecting enthusiast I'm sure you are aware of how rare it is to discover something under the ground that is in good condition and free of dirt or corrosion. It's nearly impossible. There has to be dirt in whatever you discovered buried under the ground for a an extended period. Relics and coins are dull every day that they lie hidden. In nature, precious metals such as silver and gold that resist corrosion. For the majority of time iron relics are always going to be dull and pitted.

Bronze, brass and copper will almost always are coated with a patina of some kind. The treasures we discover on a treasure hunts must be cleaned to restore their splendor. The most important question going through your head is: how can I get my treasures cleaned? At this point I believe that Google will be a great resource should you decide to. Utilizing your internet browser to find out how you can get rid of the treasure you have found is an excellent option to consider. However, I will try my best to help you understand some.

How do you get rid of Your Found Coins

I will show you how to remove the dirt from your coins after you have find them, no matter if it's copper or silver or if it is old or brand new. The method you use to clean your coins is vital. If you scrub the coins rough manner, they could be lost in value. Be aware of the lessons I'm going to share with you to ensure that the time and effort you spend looking for coins won't be an utter waste.

There are many methods that can provide effective cleaning without causing any damage to the coins. If you would like to tidy up your coins prior to selling they are, there are suitable ways to do it but I would advise you not to apply jewelry cleaner or polishing metal for your coins since these substances can be corrosive enough to harm your precious coins , rendering them ineffective. Be sure to clean your hands using soap prior to washing your coins. Do not use paper to clean a coin however, you can use a towels to clean up.

How to Clean Your Treasured Coins

* Open a faucet of water, and let the water to wash through your money. As the water

runs on the coins, dirt will be swept off. The faster the water runs, the more efficient.

The act of soaking your coins in soapy water can help remove the dirt from your coins. Simply add a small amount of detergent to the bowl of warm water and place your coins into it. It's a little bit of shaking to ensure a thorough cleaning. However, make sure you don't place a lot of coins at once in order to keep them from scratching.

* You could also utilize hot running water to clean up your coins

* You can make use of a toothpick or a soft bristled brush to remove any dirt on your coins

* Soak the coins that are dirty in a glass of white vinegar since vinegar contains acid that could help take stubborn dirt off your coins. Soak the coins for approximately 40mins to ensure an effective cleaning.

A final rinse using distillate water will add the shine to your coins. you can air dry or pat them on the soft towel. Do not scratch your coin. Pat them gently and allow the sun to take care of the rest, but in the event that the rinse was with distilled water, let them air dry, without patting.

As I've described earlier, I'm confident that you are able to clean your coins without fear. By following these steps, you will be able to take away dirt, grime and contaminants from your coins and make them sparkling.

Cleaning Jewelry

I will show you the best method to clean your jewelry and bring it shine in this article. I will teach you how to make use of everyday tools to clean necklaces, rings, earrings and other jewelry items. The most effective method to clean your jewelry after you come across them and take the jewelry home, is mixing a couple of drops liquid dish soap and several cups of water. Put the jewelry into the solution for a couple of minutes. After a few minutes you are able to take it out. Make use of a clean, dry fabric or soft sponge to clean dry and then buff off any marks. Let the jewelry dry before you contact the buyer.

How to clean up your gold

Gold is not always dirt. When you take gold, it shines as if it was not dug out from the ground. You can employ a towel for cleaning it up . The safest method to clean the gold is by soaking it in warm soapy water and then gently scrub it clean using the help of a baby brush that does not have sharp edges. Then, dry it using a gentle cloth or dry it with air, but not under the scorching sun. Avoid using ammonia or bleach to clean gold as they are extremely hard to work with.

How do you clean iron relics by using Electrolysis

Electrolysis is my method of choice to clean my relics since they contain a lot of rust inside its body. Cleaning relics is extremely difficult especially if you don't know the method for cleaning. Imagine an iron that been buried for a lengthy period of. Irons will eventually old and rusty. Cleaning it to the desired flavor will be a bit difficult however electrolysis has come to be a major factor to ease the burden particularly when it is used properly.

Electrolysis is a great option to utilize for relics that have potential worth. Electrolysis is the method that uses a direct electrical current to stimulate an otherwise unnatural chemical reaction. It is essential to utilize extremely low voltages , in short intervals.

You can purchase an electrolysis equipment, or you can build your own. Before starting, make sure that you're in a well-ventilated space and away from open flames. One of the byproducts from electrolysis are hydrogen gases.

The steps you can take to get rid of the ruins that have been lost

* You will require power supply and a meter
* Put some water into the plastic bowel, then put some baking soda in the plastic bowel

Attach to the ((+) on the thing you'd like to eliminate and then attach the (-) to the item you want to sacrifice. (+) on the object you wish to remove from the multimeter

* Put on the (+) terminal onto the item to be sacrificed, and the minimal (+) to the object that needs to be cleaned. Then, add them to the bucket.
* Turn on the electrolysis

Cleaning the item can require a few hours. Try to be sure to check it every now and then until all is working properly. After about 4 hours, the product is ready. Therefore, you need to switch off the power and take it out. Utilize a brush to cleanse it with the water in a bucket and rinse it. Relics are now ready to be used in this condition.

How to Keep Your Iron Relics

If you wish to protect your iron relics, be sure required to follow the steps I'll teach you in this article. Conserving your relics will help it last longer and appear attractive to potential buyers To do this ensure that your relics are dried with sunlight or in the oven. However, oven drying is quicker so I'll use oven.

Set the oven at 200 degrees Celsius then place the iron relics into the oven, and let them sit for at least 30 minutes before taking the relics to. Utilize a paraffin for the preservation of your relics . You can also make use of tea light candles as they are inexpensive to buy and also a an excellent material to preserve. Be sure to grease the object you intend to preserve thoroughly. While you're greasing, ensure you didn't apply pressure to it since it can be very hot in that moment. Make sure you apply enough grease it. Let the paraffin dry out of the oven, and keep in mind that drying the object at this phase will take some amount of time. After drying, scrub the surface thoroughly. In this point, your relics have been preserved.

The Rules of Metal Detecting

Every legal thing in the world is governed by rules and rules and. Our church, school and party grounds all have rules that guide them. The same is true for our individual streets. Without regulations and rules it is likely that people begin to be disruptive on the streets.

Metal detecting is a hobby where anybody can join it, however, observing guidelines and rules that govern the area you wish for hunting, is much more crucial than anything else. Metal detecting is governed by rules of behavior in England as well, and in UK I would say it is considered to be the most disciplined and well-organized activities. Metal detectors within a scheduled monument requires a permit and is punishable by imprisonment or fines if you fail to comply with. As detectorists, it is your responsibility to reach out to landowners to obtain written permission to undertake any investigation on their property. They could also agree regarding sharing what you discover if the landowner is truly interested in the subject.

1. Don't trespass on property: If you're looking for treasure, make sure you get consent from the proprietor or the occupant of your property. Remember that all land have owners. Even church, school parks, beaches, and parks have owners. Find out the owner of the property, whether it's privately owned by private people or public before going doing any treasure hunts. This is to ensure that you are not to be a target for harassment.

2. When searching for treasures, ensure that you stay clear of places where pipelines were dug up or wires could have crossed. It is essential to make sure that you do not at risk because pipelines carrying flammable liquids or gas could be dangerous.

3. Avoid national parks and state parks as well as monuments as they are prohibited

4. Hunting in a military area in which a dangerous weapon or equipment could have been dug is strictly prohibited.

5. Use a reasonable amount of caution when digging in any direction particularly in areas where you're not certain of the soil

conditions as deepseeking detectors may detect hidden cables, pipes, and other potentially hazardous materials. The appropriate authorities should be informed.

I personally request you to ask permission to enter any place you have no knowledge of prior to you discover regions. You should seek permission from the appropriate authorities.

Getting Permission for Metal Detecting

Before you begin metal detecting on private property, such as abandoned homes, old houses in different states, or even different countries, consider and get permission. At the start of this article I've listed a few of the metal detectors that be suitable for different types of metal and I'm sure by now you've got them, and have also learned how to use them. In any case, you could find metals in an abandoned area close to your home. If you are lucky hunt in a nearby zone will permit you to be aware of some rules and regulations that govern the area. There are places where you can hunt for treasure without having to worry about the owner as the location was largely ignored for many years. But , you should be cautious of dangerous animal. I suggest you take a partner in this situation. Metal detectors in private properties close to old homes is the quickest method to locate precious coins or relics to add to your collection. But first, you must obtain permission from the owner.

Find the owner

Find the owner of the property and tell him/her the things you would like to find on his or her personal property, or on the farm. If you're granted permission to access the property and locate your treasure of choice You can continue however if the property owner doesn't want you to find it the treasure, you are left with no choice but to leave.

Chapter 14: Where Can You Locate Buried Treasure?

I'm sure you've purchased the metal detector of your dreams and you're probably brimming with joy to set out and search for treasures. Also, I know that you know how the device functions and can identify the alarms and settings. In addition, I know well the type of questions are going through your head, such as where can I get these treasures (silver, gold coins, relics, coins, rings, and many more)? When I began my passion for hunting, every day I wanted to go hunting, this question has always came up in my head. I'll show you some locations where you can find valuable treasures throughout this book.

Where can I begin in metal detecting?

You can begin your hunt for metal at these locations:

* Local parks: Older or local parks are better to search for coins. Coins are often scattered throughout local parks. There are plenty of coins to hunt in parks near your home or even your neighborhood. These are great places to conduct your metal-detecting hunt as it can help you improve your hunting abilities since there is a chance to find certain coins in these old parks which have been abandoned for many years.

Make sure you get permission from the authorities before you start searching in order to avoid any confusion .When searching, ensure you maintain your detector in line with the ground, and then move it slow to allow effective detection to occur as the manner in which you place your detector can be crucial in detecting treasures.

"Old site" refers to all the abandoned houses that have owners who may not live, or might have moved out of the area due to conflict or some other reason. Old sites are a great location to find treasures. However, before you begin any type of search in this location, it is essential to do some thorough research on the location. It is possible to use the library in your area or browse the internet for historic maps of old homesteads farm, railroad camps, and farms, all of which are great for efficient searching of treasures particularly in the event that it was not discovered before.

* Sidewalk The Sidewalk is an ideal spot to begin your hobby of detecting. Choose a path which people use to start your hunt for treasure, as often, things fall out and the owners may not aware of it. By using your detector there's the possibility of finding various treasures and coins that were buried for a long time. There is a possibility that you will discover valuables along a path.

Beaches: Beaches are excellent for finding treasures. I have been to the beach to look for treasures and I'm able to declare that there's no time that I have gone to the beach without finding a certain discovery that I will take back home with me.

The beach is a great place to go for a walk, particularly when you are with your buddies. If you live near the beach there is a higher chance of you being able to spot it everyday. Beach is a place where different people gather to have fun while swimming or chatting. There's always a probability of losing the valuables that they bring along and, since they don't have detectors they are unable to locate the items. I can remember the day when that a man of a certain age was looking for his gold pendant that was purchased at a high price. This man stayed for a while looking for it. In the end the search was over, and he was fortunate that I was able to spot it the usual. When he told me what he was looking for and I decided to assist him, could think that he could be in the middle of the necklace, without even realizing that it was there. My detector spotted the necklace and he was satisfied to return back home with it. You can see that beaches are an ideal, fantastic spot that you can find after you have obtained a prior approval from the relevant authority. Certain metal detectors are designed to hunt underwater too and, if it is, you are able to search in shallow water which increases the chances of locating objects.

Abandoned buildings: Buildings that have been abandoned like abandoned homes or churches, as well as other old structures often provide homes to various relics. There is a good probability of finding high-quality treasures since different people resided in the area prior to. Make sure to explore through the area in the area of old buildings as well as under and large trees where people might have used to seek shelter in times past. You can also look inside the building where you can find hidden treasure hidden in floorboards or the walls. You might need to do some research in order to find such sites or use maps to find the area.

I've provided a few locations where to begin your metal hobby of metal detecting. As time passes you'll be able to find other places to explore metal detectors.

Are you able to earn money from Metal Detecting?

In some instances, individuals don't think they will get wealthy with the metal detector they own. Don't stop working due to metal detectors However, you can earn some money using the metal detector if you research it thoroughly and discover good spots to search.

What is the best metal detector on the market?

It is the query that people frequently are asked, "what is the best metal detector available in the present?" But in the real sense, there's no definitive answer since there are so many metal detectors available and they all work and every day people purchase a new detectors with a enhanced function.

If you're looking for ways to upgrade your metal detector, I'm talking about the best metal detector that is compatible with your budget, and can be used for searching antiques and silver, gold and jewellery, this book is the one for you. For the last 9 years,

I've been able to use a variety and types of detectors made from metal. I've tried a few, and have also recommended others to my neighbors and friends and the metal detectors worked well. What better way to introduce your kids into the world of metal detecting so that they can get involved in their spare time. Metal detecting is an exciting way to discover.

Therefore, I'll teach you how to use the best metal detectors.

1. Garrett AT Pro: It is one of the top-quality metal detector that can be used in a variety of areas of detecting different metals. Garrett AT PRO can be used for all kinds of purposes, including the search for treasures, such as silver, gold or relics as well as jewelry. Thus, taking a chance is never a bad idea.

2. Fisher F22 The Fisher F22 is a great metal detector. It's perfect for children as well as adults. It can detect metals everywhere. If you are a novice looking for it is always a good idea. They are light to carry and simple to use. It is the reason your children are able to use it.

3. Minelab Equinox 800: This metal detector is perfect for hunting relics, coins and gold. It is able to be used on all kinds of ground. The gold method is ideal for nuggets of gold. It's lightweight and submerged for as high as 10 feet. It is able to withstand any environmental. Minelab Equinox 800 has a quick recovery speed as well as a target separation. It's a solid metal detector.

4. Minelab Excalibur II If you're looking to explore diving and metal detection the metal detector is the one for you. I know a friend who utilizes it and it has worked very well. This detector made of metal is waterproof and can be used up to 200 feet deep but is much more efficient when it is on the ground. It has 17 different frequencies that can assist you in finding more treasures, coins, or jewelry, and other items that were buried in the ground. It emits distinct tones for each frequency. This is the tone for silver. is distinct from the tone of the relics. It is the best metal detector for freshwater. Also available are Garrett ACE 300, Nokta makro simplex, Fisher F75 minelab CTX 3030 whites spectra

V3i Garrett ATX, and many more. Finding any of these detectors is a great idea.

Metal Detector Accessories

Metal detecting, just like any other excellent hobby is a pastime that can be improved if you are equipped with the right equipment. There are a variety of metal detectors on the market that you can pick from. I'll list a few of the instruments that are meant to be included in the metal detector box. These tools include:

1. Pinpointer: A pinpointed metal detector is an incredible device that can be utilized by both beginners and highly experienced detectorist. It makes the search for treasure simple. Imagine digging up a huge area to find a specific metal. It takes your time and energy since you don't have any of where the treasure might be. When it comes to finding and locating treasures underground, pinpointers play an important part. If you are a metal detectorist, there are a variety of tools you must have, and one of them is the pinpointer. Pinpointer can make your search effortless. Imagine knowing precisely where the treasure is. Wouldn't that be amazing? Sure, it's extremely cool. This is

the job of Pinpointer. I remember vividly the time I was using a metal detector with the pinpointer. I felt stressed during the time since my detector did identify the presence metal. However, it was difficult to locate the right target until I purchased one. Ever since it has been enjoyable to me to perform. Finding an Pinpointer to use with your detector metal is the most suitable option. Pinpointers are like detectors but the major difference is the fact that the detector comes with large coils to help to locate the area you want deep in the ground . A Pinpointer won't go as deep into the ground but it will pinpoint the area just a centimeter from the point of the pinpointer.we have many Pinpointers available. Therefore, you can purchase one today on Amazon or any other site you can trust.

Let me provide a few options to help you to consider. There are bounty hunter metal detectors Minelab detector Teknetics metal detector the XP metal detector, the whites electronic detector, and many more. However, with a Pinpointer it is easy to identify the exact location of the treasure prior to digging. Waterproof Pinpointer metal detectors are the ideal solution when you are in water. The most modern submerged Pinpointer metal detector can operate to different depths, based what model is used. For instance it is the Garrett Metal Detector Pinpointer can work at a depth of 10 feet. Therefore, there is nothing to be too difficult.

2. Battery Metal detecting batteries are an essential component for locating treasures. Without batteries your metal detector won't operate. If there's no battery inside a detector, it also indicates that there's no metal detector. In metal detector, batteries play important role. So, getting an extra batteries for your detector is the best option to ensure that, when your detector's battery gets at a low level, you are able to have the batteries replaced.

3. Headphones: Nearly all metal detectors include headphones. The headphones are an important part of a detector made of metal because it can also help in saving the battery of your detector. A speaker can cause your detector's battery get drained. Utilizing them can assist you discover small

treasures since you will be able to see when your detector is beeping rather than in the speaker. You can buy one for yourself since the one included with many detectors do not contain a volume regulator.

4. Digger/Trowel: When it comes to metal detecting using the correct digging tool to dig the kind of ground you intend to explore is vital. When choosing which tool you want to use make sure you choose one with a light and sharp edges that can make digging easy and also less energy-intensive.

5. Metal detecting shovel: When it comes to metal detection, digging a hole is something you have to perform and take part in since treasures are usually buried in the earth. Thus, picking the right shovel can make digging a hole a breeze since not all soils are identical. The shovel is ideal for hunting relics since relics are always deep within the soil. You can choose to use a normal garden shovel, or one that is serrated on the edges, which is designed specifically for detectorists. Be sure to fill your hole once you have finished digging.

6. Bags for metal detectors: Bags for metal detectors are also necessary when you go

for a hunt. The metal detector bag is where you store the equipment you might require to hunt. The metal detector bag allows your equipment all in one location and secure.

7. Pouch: If you're detecting for treasures, it is essential to keep a place to store your finds from metal detecting in to ensure that they do not be lost. A lot of detectorists have a favourite pouch they use to keep their treasures. It's not a good idea to let those valuable coin to be thrown out of your bag to the next detectorist to discover.

How do you find the Target

* If you are digging the hole that you have your detector located you, cut a c shape into an oblong and then go in instead of digging through the entire area. If there is anything that is blocking the way then remove it using your shovel and then dig.

• Retrieve your target from the hole, or employ a probe to examine the location of your target. If you discover that the target is more deep, try to place the soil over the rolled plug.

Chapter 15: Detection Of Treasures Within River And Stream

Metal detectors in rivers and streams is enjoyable. There are jewelry, coins and several objects like old fishing knives, fishing equipment and nautical equipment. If you're a lucky one, you may find gold necklaces that may have fallen off of the wearer, rings that couples who divorced could tear off and throw into the ocean, and numerous valuable items.

A multi-purpose metal detector to do this can be a huge help to detecting within the stream and river. If you're looking for jewelry and coins that is, then hunt under bridges and underpasses. Be sure to scan the sand and water very carefully.

If you're looking for ancient relics, locate areas in the stream or river with shallow water. You can consult a historical book that you can get from a library libraries or buy an map and learn more about the stream and river that you're about to find treasures in. For those who are looking for silver and gold, be looking for areas of the river that have outcrops of quartz bearing gold rock. You could also explore all areas where there are rocks, look through the sand and thoroughly. There is a high probability of getting an undiscovered gold.

Be aware that before you go to a location to hunt, bring your cellphone and take an item box to carry.

How do you identify your Metal Detector's Finds?

Metal detection has been a well-known hobby for many years, and I am aware that a metal detector is an instrument you can utilize to find different types of treasures such as coins, gold, silver and many more. However, the metal detectors aren't complete when you aren't able to recognize your treasures accurately. I will guide you how to recognize the items you find.

Technology has advanced because they've developed metal detectors that identify what you are detecting by the sound it produces. It will emit the high and low signals. On the display that is on display in the detector you'll be able to see a certain numbers of bars glowing on the display. This is called the target recognition indicator (VDI) which indicates the type of metal that is under the ground.

How Do You Identify Coins?

Coins have been the most common object found in the metal detecting hobby it is a typical query that detectorists are always asked. What are these coins made of? I'll teach you a few methods to recognize coins. When you come across coins, but you don't have a clear idea of where the coins were used in the first place, follow these steps:

The first thing you should do is to look for the coin's denomination

How do you identify Relics

Relics date back to ancient times However, some detectorists have difficulty to determine their origins. Imagine digging up treasure and you don't know what it is. It's frustrating at times. I'll teach you how to spot the relics. It can be difficult to determine precisely the time period that the relics are or whether they are original or not. The best method to find the authenticity of your relics is by meeting an expert archaeologist or museum via forum and other platforms for social media.

How do you recognize jewelry?

If you come across jewellery such as a brooch, ring, bracelet or necklace using the aid of a detector don't believe you'll find it difficult to recognize because it's something we all often see and determining whether it's genuine or not is the most difficult thing to determine. To distinguish genuine jewelry it is necessary to look for a trademark. A single trademark could provide a good amount about the quality of the metal, the country of origin, time of production, the weight of the piece or it can be examined by a reputable jewelers.

How do you market Your Finds?

Metal detecting has come quite a ways. A lot of people enjoy discovering metal on beaches, in local parks, old schools and private property with prior authorization and detecting these locations has become an activity. However, there's an essential question that people ask themselves, especially when they discover the treasure. What is the best place to make money from my finds?

In the end, many detectorists aren't keen to market their findings because they want it be a lasting memory for them and want to share with their children what they learned when they were young. Additionally, a lot of people are willing to sell their items in order to earn some money.

Whatever you want to accomplish whatever your purpose, they're all great. For those who wish to market their items I will show the best way to do doing that. Learn through this segment.

There are a variety of ways and methods you can utilize to market anything you come across during metal-detecting.

1. Shops: Depending on the location, you could conduct an investigation of shops near your area that offer silver or gold when your finds are in one category. The stores that sell silver or gold are more than happy to purchase the item from you, however it may not be at a higher cost than you would think.

2. Selling your jewelry: If discover gold jewelry and would like to sell it, visiting the majority of shops that deal in jewelry can help you discover the right buyer fast and at an attractive price. It is also possible to offer your treasures, like jewelry to friends or family members who are interested in the

piece of jewelry. If you sell it, at least to those who you trust will earn more cash than if you sell it to a retailer.

3. Relics for sale: If your finds are relics, museums is the best spot where you can sell the items. When the items are valued they'll be thrilled to purchase them from you.

4. Online: Online sales has been a convenient method to sell items for a long time. People who do business online could easily request for it. If you go to e-bay now and start browsing on the site, you'll see sellers selling all kinds of things , including discovered finds. This includes antique coins, relics as well as gold. eBay is in existence for many years, and there are people who make cool income through it. However, if you plan on selling your items on eBay you will be charged certain costs, such as shipping however, you don't have to worry because the costs aren't too high particularly if the items are not of significant weight. Additionally, social media platforms can aid you to sell your treasures.

Signing up to an Metal Detecting Club

Figure 7: The photo of a metal detector

If you are a detectorist and you're seeking a someone to aid you in learning more regarding metal detecting or are seeking someone who you can share your experiences with, or someone to learn the proper manners for metal detecting or joining a metal detector club is a great choice to adhere to. It is essential to join a metal detecting clubs since it offers a variety of advantages, at least for those who are just beginning. If you are a novice you can join local clubs that you are able to join, such as Michigan Treasure Metal Club or you can look for closest one to where you live and sign up. If there's any local club that is in your location, you might want to think about joining an online club. Within a short period of time you'll notice an immense improvement on your metal detector skills.

Chapter 16: Of Earning Opportunities In Metal Detecting

Metal detecting isn't an inexpensive pastime. It can cost you hundreds of dollars to complete your digging gear. A good metal detector could cost between 400 to $1,500. What are your digging tools or find bags coin probes? Also, there is the cost for organizing the excavation, the fuel required for the trip, as well as other expenses. There are times when you think - wouldn't it be great to earn some return on your investment?

Treasure hunting can be an ongoing job

You may be shocked to discover that there are some who make searching for treasure buried their sole full-time occupation. Professional treasure hunters invest thousands of dollars on equipment and embark on the most extreme adventures. They can discover sunken ship wrecks as well as treasure chests buried with jewelry and gold. Some aren't as fortunate. It could take years before you find the ultimate prize. It is the reason only those who have enough resources and money are able to enjoy this type of life.

Selling the metal detectors you have found is not as easy. It is best to start by sorting the junk from the valuable. Metals that are not usable can be weighed in junkyards, and can make you a few bucks in the event that there is enough to make it worth something.

If you find gold It's recommended to get it appraised so that you know if it's worth something. If you're looking to make an easy return on it, it is possible to visit a local pawnshop and cash out the proceeds. However, if you're looking for the most price for your gold then you could sell it online , like those on popular selling sites such as Ebay. However, don't expect to make the most profit because people tend to negotiate. If your findings are substantial in quantity, you could begin to create your own website, and post your findings there so that you have control over the profits. The site will appear more professional and people will not be tempted to bargain with you. A different option would be to find localprivate collectors, as well as collectors of antique artifacts.

If you own old relics or objects that haven't been able to meet the federal requirements to be declared treasures, but have a substantial historical significance or financial value, then you are able to place it on the internet for bidding.

If you are planning to sell your own items in particular if you think that they have substantial value, you should be certain to research local business laws of your region. You might require an official permit for your business or pay tax accordingly. The best option is to locate a private buyer and not have to complete all the paperwork.

Metal detectors for sale Parts, accessories, and accessories

As per a June 11, 2011 article on The Wall Street Journal website The market for metal detectors is at a high point. The reason for the increase in selling of detectors made by metal is the rising cost for gold as well as other metals on the market. The desire of people to locate valuable metals like gold so that they are able to sell them. Moreover, metal detectors are a affordable investment for a massive gold discovery. One of the leading U.S. retailers of metal detectors saw an increase in revenue of nearly 63% in the period between 2005 and 2010.

This could be a great opportunity for many interested in consider joining the selling of metal detectors. There are numerous types of metal detectors available that are available in the U.S. one can choose from. The most well-known brands include Garret, Fisher, White and Tesoro. It is possible to contact the manufacturers if looking to sell.

Metal detectorist for hire

Every day , keys, other pair of earrings or a necklace that broke and fell off the most frequent of all , lost rings. Many people are devastated in the event that the ring they lost was sentimental or was of a high value such as engagement rings wedding rings or even heirlooms. When this happens, professional metal detectors may be able to save their wedding. If you think you've acquired enough experience in of metal detection over the years it is possible to promote your services locally as an experienced metal detectorist available for hire to find the missing items. You may be amazed at how many people could call your number.

The business opportunities associated from metal detection could be lucrative.

If you are planning to pursue the route of treasure hunting on a regular basis, you'll need the passion as well as the time and the money to engage in this kind of activity. It is only time to determine if you'll be able to find the motherlode on one of the days. However, you'll not know until you start looking.

If you have a lot of metal detectors you could decide to establish your own local rental business for metal detectors. You could also provide tutorials for novices and arrange excursions. There is also the business of buying and selling by purchasing used and functional metal detectors at a cheap price, and then selling them again to make a small profits.

If you believe that your enthusiasm for metal detection can only be described as relaxingly digging in beaches or parks, and even your own land but don't fret because there are earnings opportunities associated with it regardless of how little. Small finds that you've made over the course of months or years, like coins or rings may only amount to several hundred dollars, however, it's an investment in your pastime. Consider it as a return on the money you spent to buy your metal detector or digging equipment.

Chapter 17: Tips And Tricks In Metal Detecting,

Metal detecting is an activity that many people love. It can help develop perseverance, endurance and discipline, as well as respect for the past and the environment in addition to respecting property. Anybody, regardless of age, can benefit from metal detection. The detectors are light and portable, making it possible for children as young as seven years old, and older adults of 92 are able to continue this pastime.

The kinds of treasures you find when you use metal detectors will depend on the methods you use, the area you select, and the equipment available. With the basic metal detector it's possible to find some old jewellery, coins, and other relics. However, if you're going to hunt for the hoard, you'll need to do more than simply strolling along the beach using your detector.

The sudden change in the world of science and technology is taking the world of metal detection by storm. The introduction of microprocessor technology for metal detectors has been instrumental in revealing the countless buried objects. Anyone who wishes to stay in the know is advised to stay abreast of the most recent developments new gadgets, the newest technology and the most recent leads.

Here are some simple tips and tricks to aid you in your journey to metal detectors.

Getting started

If you've made the decision to pursue this pastime it's important to realize that just buying a metal detector will make it easy to get started. You must be familiar with specific techniques and concepts which are best learned by someone who is already taking part in this hobby. You could ask your buddies to join you on an expedition and observe the way things are carried out.

Care and handling for the detector made of metal

The components of a metal detector comprise the long handle and an electronic sensor on the lower portion. On the upper portion on the handle will be a box that houses the meter, which shows whether or not a metal has been identified. The earphones are worn to listen to the sound the machine emits, such as an alarm or clicking sound, based on the kind of metal being detected.

When the machine is turned on, keep the coil in a straight line to the ground as far as is possible. It is possible to use a technique known as "sweeping" in which you slowly move the coil between sides in a sweep motion, scanning every square inch of ground.

To maintain the metal detector device Always follow the guidelines in the user manual. As with all machines it will eventually degrade and become degraded. However, you can extend its lifespan and endurance with regular cleaning and proper storage.

Always clean your metal detector before storing it. It's often used for lengthy days in the field and subjected to numerous weathering and climate change. Fine dirt particles may accumulate inside the detector. Warm water and mild detergent are a good option to clean up the particles which have accumulated in the bottom of the. Make sure to clean the belt using the same mixture , and an ordinary brush.

Search coils are the element of the detector made of metal that is the most vulnerable to damage, therefore it is crucial to shield it. Covers for the search coil can be purchased at shops for metal detectors. These are covers made of plastic that don't interfere with the detection signal. These are extremely affordable and can help extend the lifespan of the search coil.

Storage is essential for keeping the metal detectors operating at their peak. Avoid storing in areas which are too humid or hot. Keep the detector in an area that has moderate temperatures. Extreme temperature fluctuations during storage could harm the plastic components that make up the device.

Metal detectors contain batteries. It is crucial to get rid of the batteries prior to storage in order to stop them from leaking. Leaks in batteries can cause health risks whenever you come into close contact with the battery, in particular when children are exposed to it.

Incorporating a club

You may join a metal detecting group within your area. There is probably a list online , or you could ask an area recreation facility in the region for more information. It is also possible to inquire at the local shops where you purchased the metal detector because they'll probably have a directory or even belong to an organisation that they are part of. A metal detector group could give valuable guidance and assistance for those just getting started in this type of sport. They arrange regular group excursions and can save you the hassle organizing them yourself. These groups usually share useful information regarding leads, new discoveries and solutions to current issues confronted by the metal detector community.

Another advantage of joining a metal detector company is that they are able to issue an "Detectorist card". This is an identity card that identifies you with an organization that is responsible. This card for detectorists can be extremely useful in obtaining the permission of private owners who want to search their property. It provides landowners with assurance that you're not a random looter, but are part of a lawful, responsible group consisting of metal detectorists.

Location is Important

There is an amount of detailed research needed in order to select a place which may have treasure. It is best to start by researching the area closest to you. Are there caves nearby? A region that was once inhabited by tribes or abandoned lands that were once owned by wealthy monarchs? What about a mine site fairgrounds, fairgrounds or the wreck of a ship that happened at a nearby beach? If you've studied the past of your local area and have a good understanding of the area, you might be able to identify where the hidden treasure could be. There are clues that are found in books about history as well as tales from the elderly and long-time residents who live in the region. There could also be news reports from the internet or even rumors posted on online forums that may help to add information. Some go to the point of providing leads for payment and even giving a percentage of the proceeds once the treasure has been found.

Composition of Soil

Contrary to what many believe soil conditions aren't important when you are looking for treasure. Digging in the beach is quite different from digging in the Rocky Mountains. The tools you will need to put into your bag will vary based on the area you are seeking to go to.

In addition, the properties of the soil may have an enormous role in how well the detector performs. This is due to the fact that the soil itself is magnetic. This is known as ground mineralization. Mineralized soils tend to be colored red. This is a common occurrence in soils of old that have been exposed to conditions of the surface for long periods of time. In a research study published by The Journal of ERW and Mine Action, it is claimed that the soil's electronic conductivity as well as magnetic susceptibility may influence the efficiency of detectors made from metal. The soil's conductivity can trigger false alarms, decrease the sensitivity of a metal detector, or make the detector ineffective.

Conclusion

Metal detecting is a pastime which has seen an growth in enthusiasts all over the globe. The prospecting of individuals has provided valuable historical and informational pieces that otherwise would have been in the shadows and forgotten. The equipment needed is quite simple and can be built in your home garage, if you're inclined to. It's a pastime that everyone from all walks of life are able to enjoy equally. When it first began to develop the activity the archeological community could not consider it legal. They believed that it was an affront to every ideal they believed in. The change in perspective of archeologists is something everyone should admire.

The consistent and relentless determination of a group of passionate people caused the archeological world to look up and be aware of the benefits that metal detecting could bring and how it could help everyone in the world. One thing that makes this activity an excellent hobby that deserves praise is taking great care to ensure that the recovered item is handled and stored in the most efficient possible way.

Certain archeological groups are in doubt about the methods used by the detectorists and are sometimes in awe of their preservation techniques. Due to the accumulation of rubbish, it has been a pain in the neck of archeologists to go through and sort out all of the discoveries. Every hobby or activity does not come without some setbacks, and here are a few things we should ignore in the present. The detection of metal can be enjoyable and beneficial if applied without pushing the bounds. Enjoy!